T0113467

The Orion Mystery

Articles by Robert Bauval
(appeared in *Discussions in Egyptology)*

'A Master Plan for the Three Pyramids of Giza Based on the Configuration of the Three Stars of the Belt of Orion' in *DE* 13, 1989, pp.7–18

'Investigations on the Origins of the Benben Stone: Was it an Iron Meteorite?' in *DE* 14, 1989, pp.5–17

'The Seeding of the Star Gods: A Fertility Ritual Inside Cheops's Pyramid?' in *DE* 16, 1990, pp.21–29

'Cheops's Pyramid: A New Dating Using the Latest Astronomical Data' in *DE* 26, 1993, pp.5–7

'The Upuaut Project: New Findings in the Southern Shaft in the Queen's Chamber of Cheops's Pyramid' in *DE* 27, 1993

'The Adze of Upuaut: The Opening of the Mouth Ceremony and the Northern Shafts in Cheops's Pyramid' (to appear in 1994). With A. Gilbert

'The Horizon of Khufu: A Stellar Name for Cheops's Pyramid' (to appear in 1994)

'Logistics of the Shafts in Cheops' Pyramid: A Religious Function Expressed with Geometrical Astronomy and Built-in Architecture' (to appear in 1994)

Also by Adrian Gilbert

The Cosmic Wisdom Beyond Astrology

THE ORION MYSTERY

Unlocking the Secrets of the Pyramids

ROBERT BAUVAL AND
ADRIAN GILBERT

THREE RIVERS PRESS • NEW YORK

For Michele and Dee

Published by Three Rivers Press, New York, New York.
Member of the Crown Publishing Group.

Originally published in Great Britain by William Heinemann Ltd. and in the United States by Crown Publishers in 1994.

First American paperback edition printed in 1995.

Random House, Inc. New York, Toronto, London, Sydney, Auckland
www.randomhouse.com

Printed in the United States of America

Library of Congress Cataloging-in-Publication Data
Bauval, Robert.
The orion mystery : unlocking the secrets of the Pyramids / by Robert Bauval and Adrian Gilbert.—1st American ed.
Includes bibliographical references and index.
I. Pyramids—Egypt—Miscellanea. I. Gilbert, Adrian (Adrian Geoffrey).
II. Title
DT63.5.B34 1994 932—dc20 94-19574

ISBN 978-0-517-88454-6

146122990

Contents

Plates

Acknowledgements and thanks for permission to reproduce photographs are due as follows:

Rudolf Gantenbrink for 10b, 11a and 11b; Dr I. E. S. Edwards for 6; The Royal Observatory, Edinburgh for 8; *Burham's Celestial Handbook*, Dover Publishing for 7; Dr Vagn Buchwald for 14b; Dr Brian Mason for 14a; Dee Gilbert for 1a, 2a, 9b, 10a

Line Illustrations

All illustrations are by Robin Cook

Acknowledgements

The Orion Mystery is the result of ten years' research. It is not easy to thank properly everyone who has so kindly helped to turn this project from an idea into reality. Foremost we must thank our respective wives, Michele Bauval and Dee Gilbert, for the practical and emotional support they have given us. Without their help *The Orion Mystery* would have remained a mystery for much longer.

Special thanks go to Dr I.E.S. Edwards (Keeper of the Egyptian Antiquities at the British Museum 1954–1974); Engineer and archaeologist Rudolf Gantenbrink; Professor Jean Kerisel (General-Secretary of the Franco-Egyptian Society); Professor Hermann Brück (Astronomer Royal of Scotland 1957–1975); Professor Mary Brück (Lecturer in Astronomy at Edinburgh University); Dr Jaromir Malek (Director at the Griffith Institute, Ashmolean Museum); Professor Nicolas Mann (Director of the Warburg Institute, London); Dr Vivian Davies (Keeper of Egyptian Antiquities at the British Museum); Dr Alessandra Nibbi (Egyptologist and Editor of DE Publications); Jean-Paul and Pauline Bauval (Malaga); Denis and Verena Seisun (San Diego); Marion Krause-Jach (Berlin); Lyndsay Kent (Birmingham); John and Josette Orphanidis (Cairo); Geoffrey and Therese Gauci (Sydney); Dr Henri Riad (Cairo); Robin Cook (Glastonbury);Viviane Vayssieres (London); David and Christiane Joury (Riyadh); Michael and Sue Pim (Beaconsfield); David Keys (London); Alice Harper (London); Erich Von Daniken (Solothurn); Hoda Hakim (Cairo); Osta Sabry (Cairo).

We wish to express special appreciation to Bill Hamilton and Sarah Fisher and all at A.M. Heath & Company Ltd; Sarah Hannigan and Jo Mayer and all at Heinemann Mandarin.

We would also like to thank the staffs of the various libraries that were consulted: the Mitchell Library of the University of Sydney; the Ashmolean Library, Oxford; the Warburg Institute Library, London; the Beaconsfield County Library; The British Library.

Last, but not least, a very special 'thank you' to the Robot UPUAUT 2 who 'opened the ways' for all of us.

Permissions for quotes/photographs:

We would like to thank the following individuals and organisations: Penguin Books Ltd for special permission to quote from J.B. Sellers's book, *The Death Of Gods In Ancient Egypt*; Chatto & Windus Ltd for permission to quote from E.C. Krupp's *In Search Of Ancient Astronomies*; Thames and Hudson for permission to quote from R.T. Rundle Clark's *Myth and Symbols In Ancient Egypt*; The Egypt Exploration Society for permission to reproduce Sir A. Gardiner's article in *JEA* 11, 1925; Dr Virginia Trimble for permission to reproduce her article in MIFOAWB Band 10, 1964; The Oxford University Press for permission to quote from R.O. Faulkner's *The Ancient Egyptian Pyramid Texts*; Dr I.E.S. Edwards for special permission to quote from all his published material and relevant letters to the thesis; Dr J. Malek for special permission to quote from a personal letter; Rudolf Gantenbrink for permission to use his data; The British Museum Press for permission to quote from R.O. Faulkner's *The Book Of The Dead*.

Robert G. Bauval and Adrian G. Gilbert 1993

PROLOGUE: The Last Wonder of the Ancient World

His majesty King Cheops spent all his time trying to find out the number of secret chambers of the sanctuary of Thoth so as to have the same for his own 'horizon' (pyramid) . . .

– Westcar Papyrus, Berlin Museum

As for the pyramid of Cheops, do we know everything about it, do we really know it at all? The archaeologists thought they had conclusively explored it eighty years ago, then, lo and behold, in 1945, by pure chance, the gigantic funerary boats were found intact . . .

– Georges Goyon, *Le Secret des Batisseurs des Grandes Pyramides*

In the centuries before Christ, when Alexandria was preeminent among the cities of the Greek world and its citizens were great travellers, there were seven wonders whose reputation surpassed all others and which everyone wanted to see. Six of these – the gardens of Semiramis at Babylon, the statue of Zeus at Olympia, the temple of Artemis at Ephesus, the Mausoleum at Helicarnassus, the Colossus of Rhodes and the Pharos lighthouse at Alexandria itself – have disappeared. Only one remains for us to visit: the pyramids of Egypt.

These extraordinary monuments, which make Stonehenge look like a morning's work,[1] have inspired awe through the centuries. Their sheer size sets them apart, let alone the perfection of their geometry. Just how they were built remains a mystery; even today we would be hard pressed to replicate them with all the advantages of modern technology. At the time of the Ancient Egyptians there were no dump-trucks or cranes, no steel cables or hoists, not even iron tools. Without the benefit of so much as a simple pulley, they built mountains from stone and, with a precision that is truly astonishing, laid these out on the desert floor. Yet the more puzzling question is why and not how they built them. Why did the Egyptians choose to build pyramids when, so far as we know, they had never been built before?[2] Why did they build them so big and of such precision? Why did they scatter them around the desert instead of building them all in one place?[3]

Contemporary Egyptology has no convincing answers. Pick up any textbook on the subject and you will encounter the same statement, that the pyramids functioned as royal tombs. But why, when a simple hole in the ground would have sufficed, should the Egyptians have built tombs up to 147 metres high? Why make this prodigious effort to house a dead body? Even given that the pharaohs were autocrats and were revered as living gods, this seems like a colossal waste of time and energy.

The popular image of gangs of slaves forced to carry out this enormous task is also a myth; there is no evidence to suggest that people were compelled to take part in this massive enterprise against their wills – indeed, if anything, the opposite. The sheer quality of craftsmanship in the construction of the pyramids suggests a pride in the work, and there are subtleties of design which suggest ideals at odds with the brutal image of Ancient Egypt portrayed in biblical film epics.

In fact, the Egyptians were highly civilised and deeply religious at a time when Europeans were still primitive, and there is much to suggest that they built pyramids more as an affirmation of their religious convictions than to glorify dead pharaohs, however powerful. But the Egyptians were also an extremely reserved people, who kept the inner mysteries of their religion from all but a few chosen initiates. As it was these few who directed the building of the pyramids, it is not surprising that we know so little about their motives.

There are also mysteries surrounding specific pyramids, especially the Great Pyramid of Giza. Having stood intact for several millennia, it was first broken into in AD820 by a team of Arab workmen on the orders of Caliph Ma'moun, son of the legendary Haroun al Rashid.[4] After weeks of tunnelling through solid limestone, they emerged into a dark, gloomy passageway. Further exploration along tunnels and galleries revealed a system of three chambers which, much to their chagrin, were all empty. Only a lidless, granite sarcophagus was found in the so-called King's Chamber.

The Ancient Egyptians were themselves remarkably silent about the pyramids. By the time of Tutankhamun (*c.* 1300BC), the Giza pyramids were over one thousand years old, and the memory of who built them and why was lost. The Greeks and the Romans who occupied Egypt from the fourth century BC to the seventh century AD took little interest in these monuments, though the Greek historian, Herodotus, who spent some time in Egypt in the fifth century BC, sought to explain their origins and purpose in *The Histories*. This is the earliest first-hand account of the pyramids known to us and is a mixture of personal bias, local gossip and mythology.[5] It was not until the Arabs invaded Egypt in the seventh century AD that a real attempt was made to explore the pyramids.

The Great Pyramid has continued to fascinate adventurers and has attracted more attention than any other single building in history. Throughout the centuries there has been the suspicion that it held further secrets, that somewhere inside was a hidden chamber, and that one day this chamber would be found. Generations of Egyptologists and amateurs have searched for it, and have used everything from dynamite to x-rays, but without success.

On 22 March 1993 the international media[6] excitedly announced that Rudolf Gantenbrink, an unknown German robotics engineer, had made the most significant archaeological discovery of the decade. Employed by the German Archaeological Institute in Cairo to find a way of improving the ventilation in the Great Pyramid, Gantenbrink had sent a tiny remote-controlled robot, UPUAUT 2 ('Opener of the Ways' in Ancient Egyptian), up the southern shaft of the Queen's Chamber. Coming to a halt after about sixty-five metres, the robot sent back video pictures of what appeared to be a small door, with a tantalising gap underneath it.

Now a door suggests something beyond it, perhaps a chamber. If such a chamber exists, it could not have been plundered since the pyramid was built, as the shaft was closed at both ends. This means that whatever the Ancient Egyptians might have put in it has lain undisturbed for at least 4400 years and must still be there, and if the pyramid builders took so much trouble to conceal it, it must have been very important; more important perhaps than the mummy of a dead pharaoh. This suggests that it was something they regarded as central to their religion and perhaps connected with their motivation for building the pyramids in the first place . . .

But Rudolf Gantenbrink was not the only person interested in the shafts, for I had been investigating them for several years in connection with their astronomical bearings. By extraordinary coincidence, Adrian Gilbert and I had been taking photographs of the Queen's Chamber and the opening of the southern shaft just days before UPUAUT 2 went on its epic journey, and I met and talked to Rudolf and his team as they prepared for the final stages of their investigation into the Queen's shaft.

Adrian's and my interest was altogether more abstract: what might these shafts have symbolised? It is by now well known that they were not primarily for ventilation. It is the direction in which they point that is most significant – towards specific stellar regions which had great importance for the Ancient Egyptians. I had been researching the matter of the lost star religion of the pyramid builders for a number of years and had published several articles on the subject;[7] however, it seemed to me that some of the data on the angle of one of these shafts was inaccurate. I was therefore hoping that Gantenbrink's new measurements by laser beams would provide us with a more accurate reading, so that the astronomical target of this shaft could be verified.

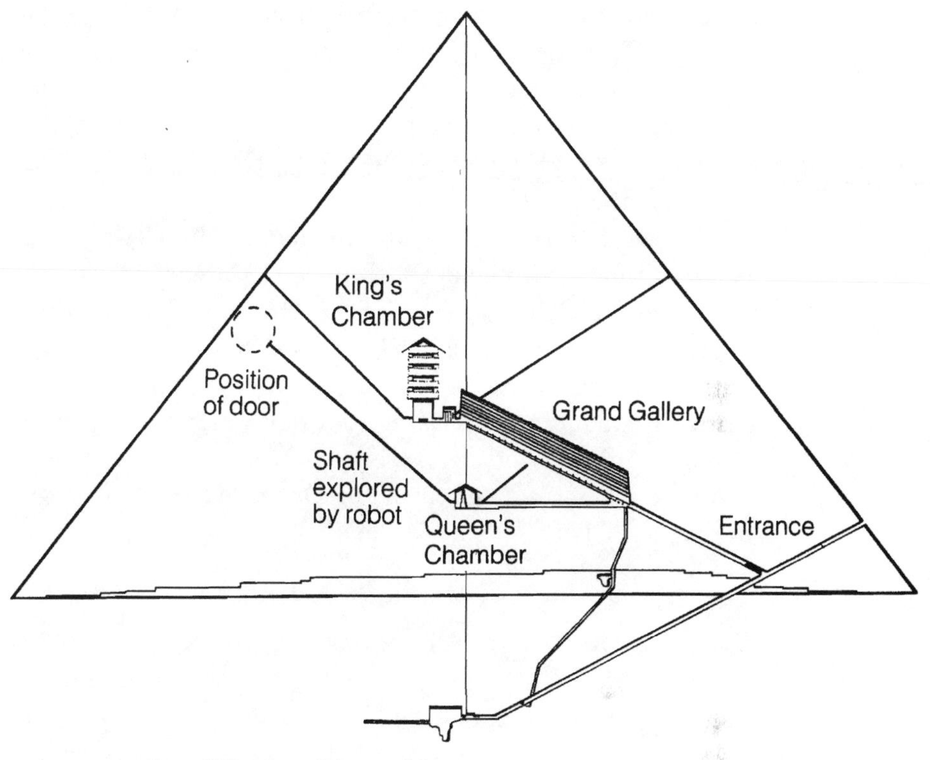

1. *Cross-section of the Great Pyramid showing chambers, passage-ways and shafts*

Gantenbrink's amazing discovery was reported on the front page of the *Independent* in London and, as his spokesman in England, I was asked by their archaeological correspondent to comment on the religious significance of the shafts. I explained that the southern shaft of the King's Chamber pointed towards the Belt of Orion, associated with the god Osiris, and the equivalent shaft from the Queen's Chamber (the one blocked by the 'door') pointed towards Sirius, the star of the goddess Isis.[8] These alignments were not accidental but were clearly bound up with the purpose of the pyramid.

This was the first the world knew of an academic debate

concerning a star religion linked to the pyramids, because the standard textbooks had always supported a 'solar hypothesis'. Speaking on Channel 4 television that evening, Dr Edwards, the world authority on pyramids, lent support to my theory by suggesting that the door might hide a statue of the pharaoh 'staring out in the direction of Orion'. On the subject of the shafts, he was quoted the next day by the *Daily Mail* as saying that 'They were called ventilation shafts because nobody knew any better. . . . They point at the constellation of Orion, whose stars were the god Osiris.'[9]

What other secrets do the Giza Pyramids hold relating to the stars of Orion? With confirmation from Gantenbrink that the true angle of the shaft, verified by UPUAUT 2, fitted exactly with my predictions, I had the final evidence that a master plan governed the building of the pyramids – simple but with astounding implications for our understanding of the Pyramid Age.

My search for a solution to the Orion Mystery had begun twelve years ago.

THE GENESIS OF THE ORION MYSTERY

1

The skies have been the mover of [man's] science for millennia, they are his hopes and dreams of tomorrow; nowhere is the vision of the first men who carved their thoughts on stone so fully displayed as in the tombs of earliest Egypt

– Jane B. Sellers, *The Deaths of Gods in Ancient Egypt*

I A Dance for Sirius

In 1979, at London-Heathrow airport, I bought a book called *The Sirius Mystery* by Robert Temple.[1] I took it with me to the Sudan, where I was going to work on an engineering scheme to connect the Blue Nile with the Rahad River by means of a canal system.[2]

The book turned out to be a historical detective story, interesting because its initial point of focus was an African tribe, the Dogon, who every sixty years enacted a ceremony called the Sigui, during which their priests put on masks and performed a complex dance. This was a renewal ceremony, based on the apparent motion of Sirius, known to most people as the 'dog star'. Sirius is the brightest star in the heavens and is in the constellation of Canis Major just below Orion.[3]

The Sirius Mystery also explored aspects of Ancient Egyptian astronomy, and as I was both an amateur Egyptologist and a keen student of Ancient Egyptian astronomy, it seemed like a good book to take to the Sudan, where the night skies are ideal for star watching.

I discovered that Temple's mystery was based on an article written in the 1950s by two French anthropologists, Griaule and Dieterlen.[4] They had studied the Dogon and found them to be in possession of unexpected knowledge concerning Sirius and its invisible partner, the 'white dwarf', Sirius B. Robert Temple, an American living in Britain, a Fellow of the Royal Astronomical Society and a graduate in Oriental Studies and Sanskrit, came across their work in the early 1960s. He was baffled as to how the Dogon could have known of the existence of Sirius B, given that it is barely visible using a very powerful telescope (it was only in 1970 that the first photograph of Sirius B was obtained with great difficulty by the astronomer Irving Lindenblad).[5] Most people today remain ignorant of the existence of Sirius B and not many would even be aware of Sirius A, so how could the Dogon have had accurate information concerning Sirius B in the 1950s?

A further mystery was how the Dogon seemed to have kept physical records relating to this star, in the form of cult masks, some of which are centuries old and are stored in caves. Their obsession with this tiny star was strange: where had their knowledge originated?

Temple concluded that as it clearly had not come from modern astronomers, it must have originated from ancestral sources and had probably been passed down to the Dogon before they migrated to their present home, Mali in sub-Saharan Africa. In Egypt, in ancient times, Sirius was considered the most important star in the sky and was identified with the Egyptians' favourite goddess, Isis. In this oblique way Temple's initial study of the article by the French anthropologists had led him via an obscure African tribe inevitably to Ancient Egypt. He wrote:

> When I began writing this book in earnest in 1967, the entire question was framed in terms of an African tribe named the Dogon. . . . The Dogon were in possession of information concerning the system of the star Sirius which was so incredible that I felt impelled to research the material. The result, in 1974, seven years later, is that I have been able to show that the information which the Dogon possess is really more than five thousand years old and was possessed by the Ancient Egyptians in the pre-dynastic times before 3200BC.[6]

Though much of the rest of his book concerning the mythology of the Near East was highly speculative, Temple had uncovered a mystery worthy of further investigation. If the Dogon had inherited their knowledge of Sirius B from the Ancient Egyptians, what other knowledge might these ancients have had concerning the stars? I had always understood that Egyptians of all periods venerated not so much the stars but the sun god, Ra; I also knew that for a short time under the pharaoh Akhenaten (*c.* 1350BC) there was some heresy concerning the god Aten, symbolised by the solar disc.[7]

In any event, at the time I read *The Sirius Mystery* I knew very little of their more ancient star religion. This subject turned out to be an interesting and neglected field of study and one of the most important in understanding the Ancient Egyptians' sky religion. It also became clear that there was so little written about it because it was esoteric knowledge of the highest order. The Egyptians were probably the greatest astronomers of the ancient world, but, unlike the Greeks and Romans, most of their knowledge was restricted to a small group of initiates.[8] At least some of these secrets concerned the stars.

It seemed obvious to me that the place to look for evidence of this lost knowledge was not among the tribes of Mali but in Egypt itself. There the ancients had left a wealth of contemporaneous evidence in the form of temples, tombs, obelisks, inscriptions and – above all – the pyramids. I had an intuition that the lost knowledge might be something of great significance, and felt a strong urge to pursue the trail.

Having finished my contract in the Sudan in 1980, I left for another engineering assignment, this time in the desert kingdom of Saudi Arabia. Little did I know that in less than a year I would come across further amazing evidence that would reawaken my interest in the star mystery and point towards a connection with the pyramids.

But before going into this, let me just review current knowledge of the Pyramid Age and what is known of the sky religion of the Egyptians of that period.

II The Land of God Kings

The land of Egypt might have been just an extension of the Sahara desert were it not for the world's longest river, the Nile. This mighty artery, with its sources deep in the heartlands of Africa, is fed by the reservoirs of Lake Tana in Ethiopia and Lakes Albert and Victoria in Uganda, and brings life to the otherwise torrid regions of Sudan and Egypt. Seen from the air, it looks like a gigantic snake, lazily slinking northwards to the cool Mediterranean. It has a presence and beauty that contrasts strangely with the burning desert beyond its banks.

The Egyptians had good reason to worship the Nile, which they believed was the manifestation of the gods. With minimal amounts of rain falling on their land, it was their only steady source of fresh water. Their lives were to a large extent governed by the rhythm of the Nile; the annual flood, caused by the melting of snow high in the mountains of Ethiopia, which occurs around the time of the summer solstice, was the most important event in their calendar. The Nile irrigated a wide area on either side of its course and deposited large amounts of thick black silt which increased the fertility of the land. So rich was this natural fertilisation that several crops were possible in a year. To all intents and purposes, Egypt was (and indeed still is) the 'gift of the Nile'.

Geographically, Egypt's inhabitable land (excluding a few desert oases) falls into two distinct areas: the long, narrow valley of the Nile as it winds through canyons and desert, and

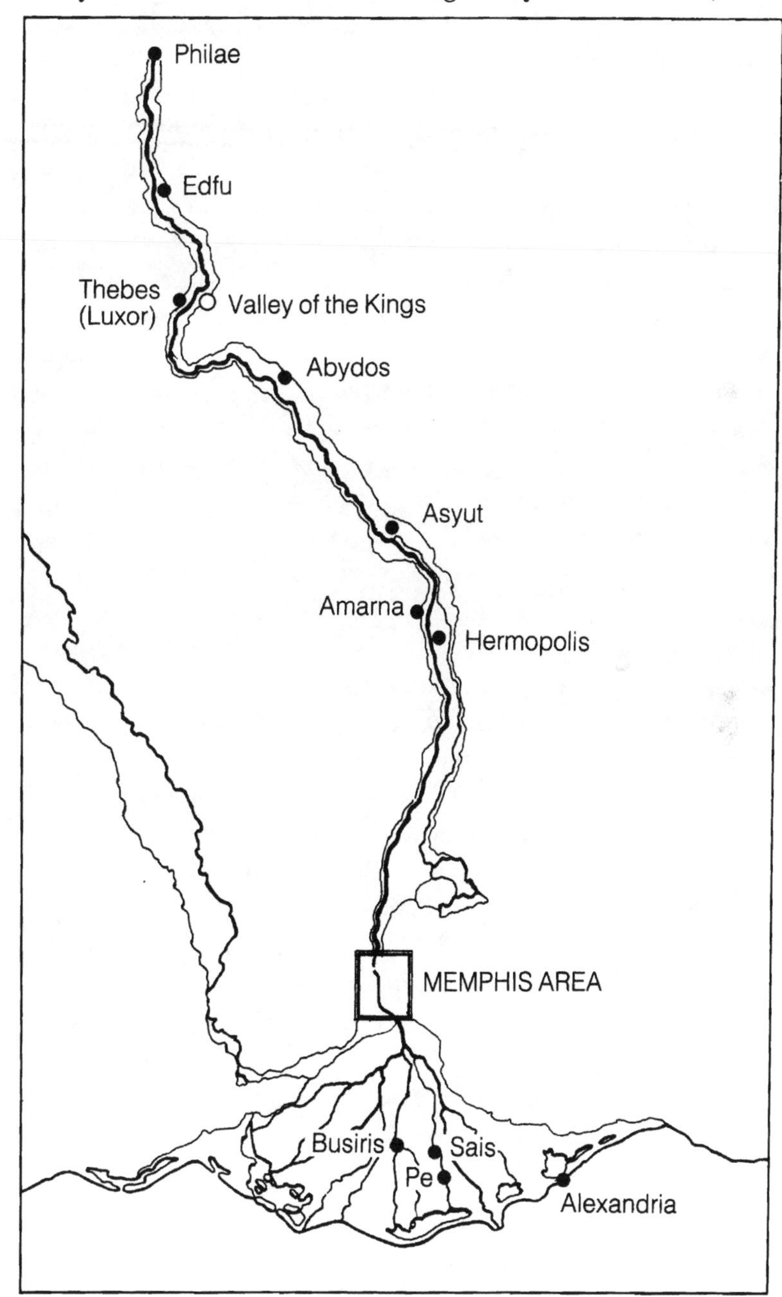

2. Map of Egypt looking south along the Nile Valley

the triangular, flat Delta where the river meets the Mediterranean. These two territories have always been referred to as Upper and Lower Egypt and are quite different in character. The fertile valley of Upper Egypt is a thin streak of land some 600 miles long and only about three miles wide. In the past this was sufficient to support the local population, but agriculture was only part of their *modus vivendi*. The Nile was also a great highway linking darkest Africa with Lower Egypt. The cities of Upper Egypt were important trading posts on this highway, dealing in ivory, precious stones, wood, incense and slaves, and this trade, as much as agriculture, constituted the wealth of Upper Egypt. Lower Egypt, by contrast, is a flat alluvial plain, with some of the finest arable land in the world, irrigated by a constant water supply. Once wholly marshland, the Egyptians transformed it into farmland. It is now an area of enormous date groves and, under the shade of the palms, other crops thrive to feed both man and animals. Rich in wheat and corn, it was one of the great food baskets of the ancient world.

The natural division of the land gave rise to the two separate kingdoms of Upper and Lower Egypt. The capital of Upper Egypt was at Nekheb, near Hierakonopolis, under the protection of the vulture goddess Nekhebet. Lower Egypt's capital was at Pe, a town in the Delta which the Greeks later called Buto; it was protected by the cobra goddess Edjo. How the two kingdoms related to each other in pre-dynastic times is not known, but Egyptologists believe that they were first united by Menes, a powerful king of Upper Egypt, who was also known as 'the scorpion king'. Around 3100BC he is said to have subdued Lower Egypt, declared himself ruler over a united kingdom and founded the First Dynasty. This date is usually taken as being the start of Egyptian history, though the Egyptians considered their civilisation to be very much older and looked back to a golden age when the two lands were ruled by the gods. They believed the anthropomorphic deity Osiris was the first divine pharaoh. The fact that Egypt was really two kingdoms was, however, never forgotten and the pharaohs were always referred to as 'Lord of the Two Lands' or 'King of Upper and Lower Egypt'. They also adopted both protective goddesses, Edjo and Nekhebet, and frequently wore a double crown, red for Lower Egypt and white for Upper Egypt, to symbolise their lordship over both lands.

Following Menes's conquest of Lower Egypt, and unification of the kingdoms, there were to be some thirty-two dynasties up to and including the Greek Ptolemies, who took control following Egypt's conquest by Alexander the Great in 332BC. Cleopatra was the last pharaonic ruler of Egypt before it fell to the might of Rome in 30BC, when the line of pharaohs effectively ended. This long history, from *c.* 3100BC to 30BC, is sub-divided for archaeological purposes into a number of periods, each comprising a number of dynasties.[9] For our purposes, the most important of these is the Old Kingdom, also known as the Pyramid Age. It comprises Dynasties Three to Six (*c.* 2686–2181BC). This period, according to Dr Edwards, is the 'Pyramid Age par excellence', and reached its apotheosis in the Fourth Dynasty during which the greatest pyramids were built.[10]

Menes built his new capital at Memphis, amid lush palm groves on the west bank of the Nile. Its location was of great political and symbolic importance for it stood near the head of the Delta, at the junction of Upper and Lower Egypt. Almost nothing now remains of this once great city: donkeys and cattle graze under the date palms where palaces and temples once stood. There are plans to carry out major excavations of the site when funds allow, but present archaeological knowledge concerning this ancient metropolis is surprisingly scant. In later times, during the New Kingdom (*c.* 1450BC) the capital of united Egypt was moved to Thebes in Upper Egypt, but Memphis continued to prosper until well into the second century AD.

A few kilometres west of Memphis is the ancient necropolis of Saqqara, a royal cemetery important throughout Egyptian history. It is the site of the famous step-pyramid of Zoser and several other smaller pyramids, notably that of Unas, last pharaoh of the Fifth Dynasty. Saqqara probably takes its name from Sokar, a falcon-headed deity believed to be the keeper of the whole necropolis.[11] It contains many other tombs, some of which are beautifully decorated with scenes of everyday life as it must have been when the pyramids were built, and exciting discoveries continue to be made in this necropolis, most recently the tomb of a general of Ra-Moses II, north of Saqqara. Unfortunately, like the pyramids themselves, these tombs are suffering from the depredations of tourism, and are urgently in need of protection to prevent further deterioration.

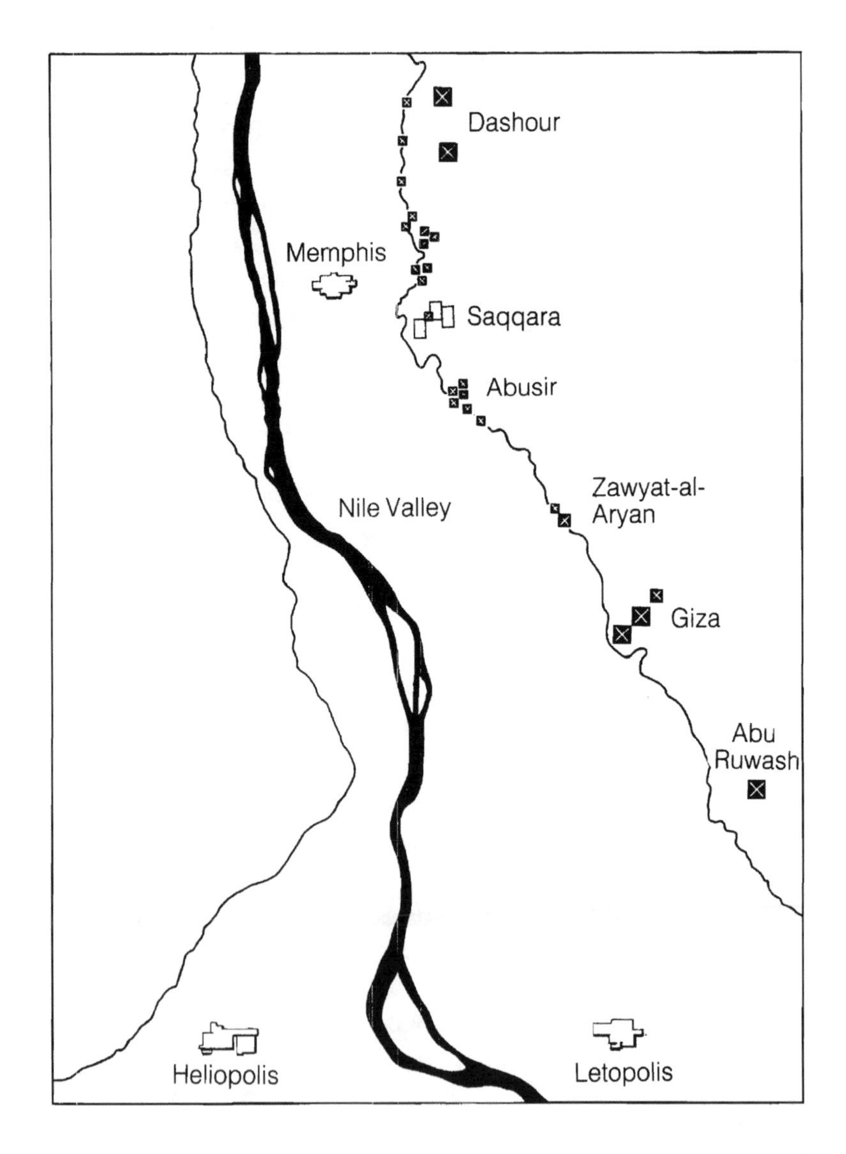

3. Detail map of Memphis area

On the other side of the river from Memphis and some twenty kilometres north is the legendary sacred city of Annu or Heliopolis,[12] as it was later called by the Greeks. This was the seat of a powerful priesthood whose members were the custodians of a school of wisdom or initiation, and the great temple of Ra, the sun god. The priests of Heliopolis wielded enormous influence as custodians of the state cult; their school of wisdom was famous well into the Ptolemaic period, and is mentioned with great reverence by Herodotus.[13]

Heliopolis is now a thriving suburb of Greater Cairo and little remains of its great past; only an obelisk of Sesostris I, a powerful pharaoh of the Twelfth Dynasty (*c.* 1940BC) and a few broken pillars and beams of an ancient temple. The Sesostris obelisk stands in lonely isolation resembling a great stone finger pointing to the sky. Yet it is only one of many that once stood in Heliopolis, including two raised by Tothmoses III of the powerful Eighteenth Dynasty. These were moved by the Romans to Alexandria in about 12BC and placed in front of the Caesarion, a temple dedicated to Augustus Caesar. Neither Tothmoses nor the Romans could have imagined that several millennia later, in AD1878, these two obelisks would leave Egypt altogether. One now stands on the Victoria Embankment in London and is known, erroneously, as Cleopatra's Needle; the other is in New York's Central Park outside the Metropolitan Museum of Fine Arts.[14]

On the edge of the desert, across the river from Heliopolis and a short way upstream, is the elevated plateau of Giza, known to Egyptologists as the Mokattam Formation. This site, now almost engulfed by the spread of Greater Cairo, is where the famous trio of pyramids, the last Wonder of the World, proclaim the glory of those who built them: three of the great Fourth Dynasty pharaohs whom the Greeks called Cheops, Chephren and Mycerinos.[15] At Giza there are also much smaller 'satellite' pyramids, arrays of small, flat tombs, remnants of chapels and temples and, of course, the legendary Great Sphinx.

It is the Giza pyramids which have excited the imagination of generations and are what most people think of when they hear the word 'pyramid'. Few modern visitors are aware, however, that the father of Cheops built two other giant pyramids at Dashour, about twenty kilometres south of Giza. Unfortunately, the army controls the site of Dashour, and it is out of bounds to the regular tourist. In the twenty-kilometre stretch from Dashour to Giza are the pyramid 'fields' of Saqqara, Abusir and Zawyat Al Aryan, and about six kilometres north-west of Giza is the desolate site of Abu Ruwash, where a pyramid of the Fourth Dynasty once stood. Only its base and foundations now remain. Meidum is about sixty-five kilometres south of Saqqara, but is generally not considered an integral part of the Memphite Necropolis, which covers an area thirty kilometres long by four kilometres wide.[16]

About seventeen kilometres north of Giza and on almost the same latitude as Heliopolis was another important centre: the ancient city of Khem, later called Letopolis. This Delta city was closely connected with the hawk god Horus, and a very ancient temple, older than the pyramids, was sited there.[17]

III Heliopolis and the Temple of the Phoenix

At the time the pyramids were built, there were no obelisks at Heliopolis, only a crude sacred pillar from which, apparently, the ancient name of the city, Annu, derives.[18] Cairo did not exist and Heliopolis was the religious heart of the country. At Annu stood a temple dedicated to Atum, the Complete One, the father of the gods. During the Pyramid Age Atum became more and more identified with the sun god, Ra, who in time usurped Atum's place and demoted him to the role of 'old sun' as it sets in the west. However, in these early days, before the Pyramid Age, Atum stood for the One God, equal to our concept of God the Father. Atum was the creative power behind the sun and everything else in the world.[19]

At Heliopolis there was an important sacred hill or mound upon which the First Sunrise had taken place,[20] and belief has it that the sacred pillar stood on this holy mound prior to the Pyramid Age. At the beginning of the Pyramid Age, another, even more sacred relic either replaced the sacred pillar or, more likely, was placed upon it.[21] This was the Benben, a mysterious conical stone which, for reasons we will discuss later, was credited with cosmic origins. The Benben Stone was housed in the Temple of the Phoenix and was symbolic of this legendary cosmic bird of regeneration, rebirth and calendrical cycles. In Ancient Egyptian art the phoenix was usually depicted as a grey heron, perhaps because of the heron's migratory habits; it was believed that the phoenix came to Heliopolis to mark important cycles and the birth of a new age.[22] Its first coming seems to have produced the cult of the Benben Stone, probably considered the divine 'seed' of the prodigal cosmic bird. This idea (which we shall also consider later) is evident from the root word *ben* or *benben* which can mean human sperm, human ejaculation or the seeding of a womb.[23] The mysterious Benben Stone disappeared long before Herodotus visited Egypt but not before it had bequeathed its name to the apex stone or pyramidion usually placed on top of pyramids and, later, the head of an obelisk.[24]

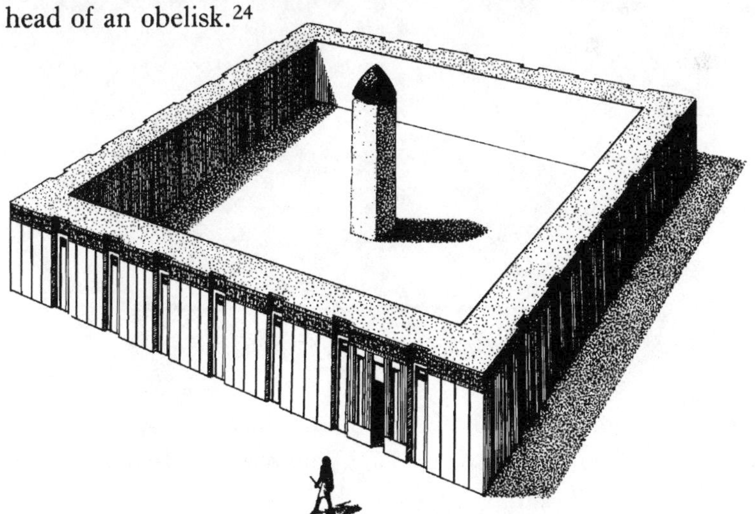

4. Artist's impression of the original temple of the Phoenix at Heliopolis, with Pillar of Atum surmounted by the Benben stone

What was the Benben and what became of it? It was clearly at the centre of an important royal cult which later

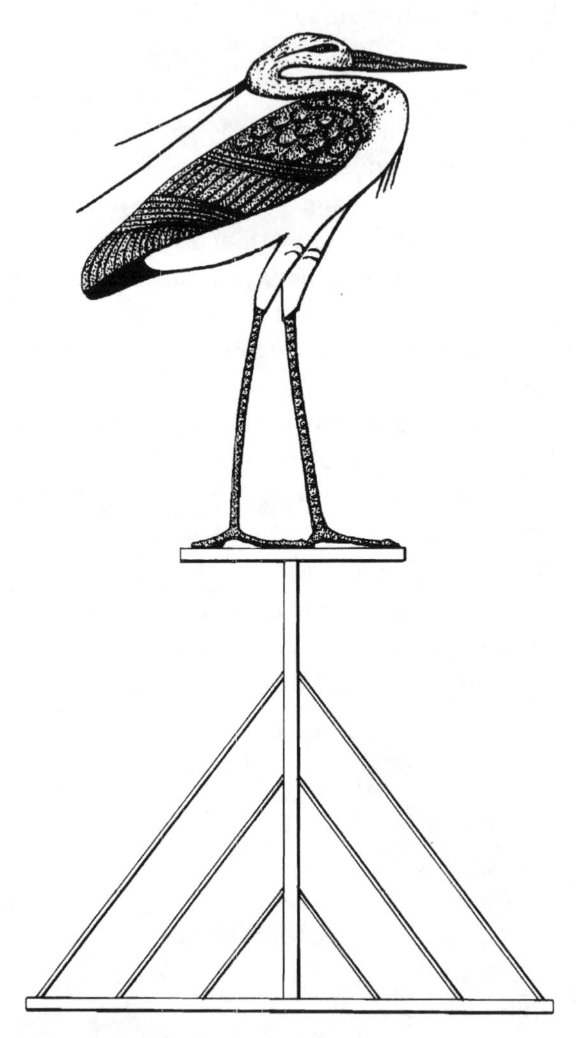

5. *The Egyptian Phoenix or Bennu Bird*

built pyramids. As we have said, the holy city of Heliopolis was in the hands of a priesthood which wielded considerable power in the Pyramid Age, and there can be little doubt that the design of the pyramids was under their direction.[25] The word priest as we understand today it is somewhat misleading, for the Heliopolitan sages were most likely highly trained initiates conversant not only with religious ideologies but with the study of celestial bodies and, probably, the art of symbolic architecture and hieroglyphs, the sacred form of writing invented by the Egyptians.[26] Clearly, then, the Heliopolitan priesthood would have known about the mysterious stellar religion alluded to in the *Sirius Mystery*.

Egyptologists consider that Heliopolis provided the nearest thing to a state cult, for while every district had its own local gods, the Heliopolitan religion, whose pantheon was the Great Ennead of gods, was recognised everywhere.[27] This great pantheon, composed of nine deities, formed the family ruled by Atum-Ra. Originally un-manifest, Atum, or Atum-Ra, masturbated and thus created Shu, the air god, and Tefnut, the moisture goddess. This couple created Geb, the earth god, and Nut, the sky goddess. Geb and Nut mated, though their copulation was interrupted by their father Shu, who, as the air, came between them and lifted the canopy of the sky away from the earth, thereby parting the divine lovers. In spite of this *coitus interruptus*, Nut, the sky goddess, gave birth to four anthropomorphic gods who lived on earth. These were Osiris and Seth, two male gods, and their sisters, Isis and Nephthys. Osiris and Isis were united and became the subject of Ancient Egypt's greatest myth as the first divine couple who ruled Egypt. Isis gave birth to an only son, Horus, from the seed of Osiris. Since Osiris, or his 'soul', was often identified with the phoenix, it is probable that the Benben Stone symbolised, among other things, his seed, and thus the generative power that created Horus from the womb of Isis.[28]

It is with these last five anthropomorphic or human-form gods that we shall be mostly involved, and especially with Osiris, for he was not only seen as the first divine king of Egypt, but his tragic death and miraculous resurrection provided the basis of the ancient Egyptian mysteries and the origin of their rebirth cult.[29]

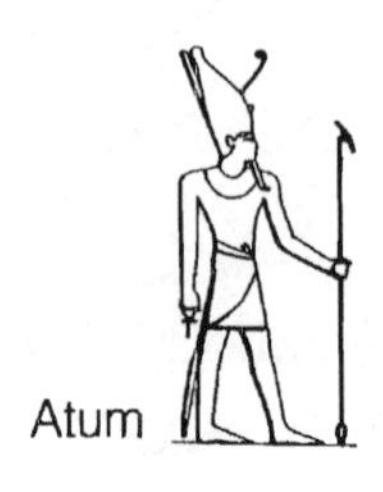

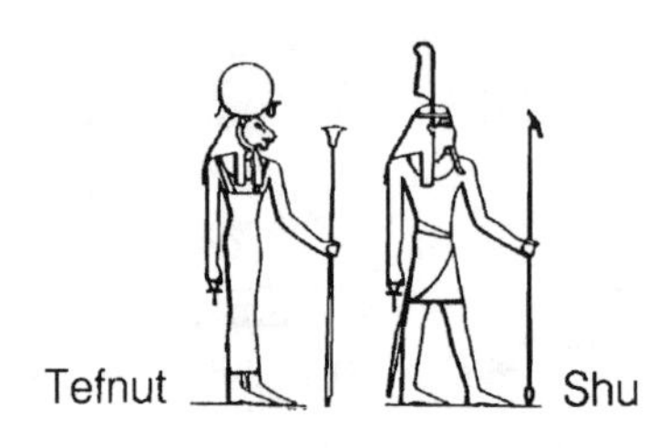

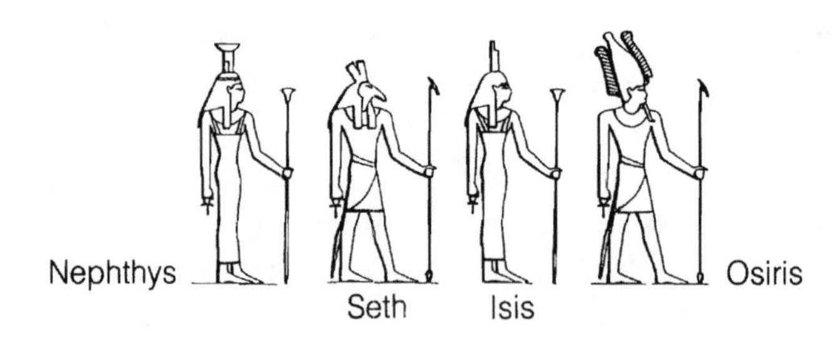

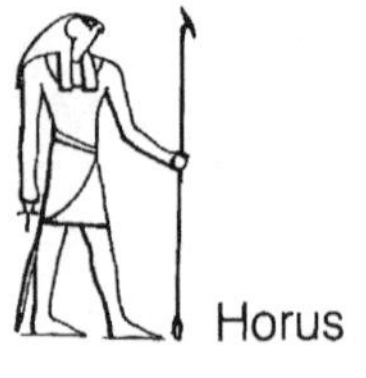

6. *The* Heliopolitan Ennead *of* Gods *plus* Horus *the son of* Isis *and* Osiris

IV The Pyramid Age

The Ancient Egyptians were religious people and believed emphatically in an afterlife in some heavenly Egypt. To help the dead reach this celestial afterworld, it was deemed important to preserve the body of the deceased as far as possible and to provide the departed with the means and accessories for the arduous journey into eternity.

In pre-dynastic times the dead were buried in simple pits dug in the desert sand. The body was placed on its side in a foetal position, presumably ready to await a rebirth in the afterlife. In the dry conditions of Egypt's western desert, natural mummification took place, probably more by accident than design. Yet the corpse was always liable to be exposed by jackals and wild dogs, desecrating it and making it easy for robbers to locate the tomb and steal precious artefacts. During the First Dynasty the Egyptians began building tombs with superstructures of mud bricks and stone to cover the burial pit and protect the corpse. These were massive rectangular structures with flat roofs commonly known as *mastabas*.[30] There are many of them scattered in the Memphite Necropolis. They continued to be used throughout the Old Kingdom period. Up to the end of the Second Dynasty some kings were buried in them, but thereafter they were used only for the nobility; dead kings were to have much grander 'Mansions of Eternity'.[31] Mastabas were made of rock and mud bricks and are believed to have been more elaborate than the homes in which the people themselves lived. According to Dr Edwards, the reason for this was religious:

> In a land where stone of excellent quality could be obtained in abundance, it may seem strange that the rulers and governing classes should have been content to spend their lives in buildings of inferior quality to their tombs. The Ancient Egyptian, however, took a different view; his house or palace was built to last for only a limited number of years . . . but his tomb, which he called his 'castle of eternity', was designed to last for ever.[32]

During the Third Dynasty the so-called step-pyramids appeared but these are not true pyramids in the geometrical sense of the word, so we would do better to think of them as stepped towers.[33] The largest of these remaining is King Zoser's at Saqqara, not only the largest of its kind built in Egypt but the first known structure built with stone masonry: quarried and properly cut rather than rough stones stacked together.[34] This innovation is attributed to a genius priest-architect called Imhotep, who was also Zoser's vizier. Imhotep, whom the Greeks equated with their god of medicine Asclepius, was later deified and said to be the greatest wise man. He was also the high priest of Annu and astronomer-general or chief stargazer, with the title 'Chief of the Observers'.[35]

The step-pyramid of Zoser is an imposing structure; it is sixty metres high and has a rectangular base. It seems to have contained the burial chamber of King Zoser and others used for his family. Its ziggurat-like structure seems also to have symbolised a ladder whose six steps leading up to a seventh platform probably corresponded to the planetary spheres which encircle the earth and therefore to stages of ascent through which the soul must pass after death. This is a common concept found in mythologies around the world and is well documented in William Lethaby's *Architecture, Mysticism and Myth*. Speaking of the ziggurat of Borsippa, restored by Nebuchadnezzar, Lethaby translates the latter's inscription:

> I have repaired and perfected the marvel of Borsippa, the temple of the seven spheres of the world. I have erected it in bricks which I have covered with copper. I have covered with zones, alternately of marble and other precious stones, the sanctuary of God.[36]

Writing of the step-monuments of Egypt after several pages considering similar structures in Assyria, China, and Mexico, Lethaby says:

Maspero and Perrot are disposed to accept the account of a Greek writer that the Great Pyramid was decorated in zones of colour, with the apex gilt; and it would seem more than a coincidence that the earliest pyramids, attributed to the first four dynasties, should be in stages. That at Sakkarah still has six steps, decreasing from thirty-eight feet high at the bottom to twenty-nine feet for the top, in remarkable resemblance to the Ziggurat at Babel.

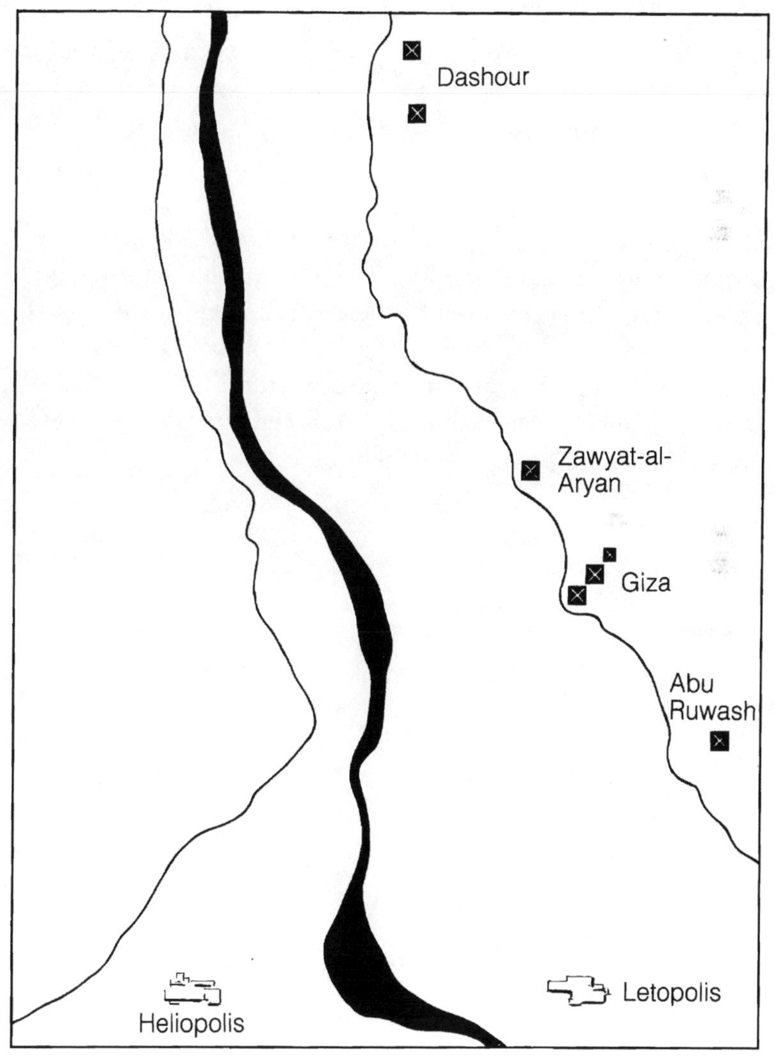

7. *The Pyramids of the IVth Dynasty*

> And Mr Petrie has found that the pyramid of Medum [sic] was built in seven degrees before the outer and continuous casing was applied, 'producing a pyramid which served as a model to future sovereigns'.[37]

Looked at in this way, the step-monument of Zoser is much more than the tomb of a powerful king; it is a statement of religious beliefs and an expression of the highest art. It stands proudly above Saqqara amid the tombs of generations as a symbol of the Egyptian religion. Visible from Memphis and the surrounding Nile Valley, it would have been a constant reminder that the purpose of life on earth was to prepare for the hereafter.

Following Imhotep's achievement at Saqqara, several other step-pyramids were built, notably at Meidum, some forty-five kilometres south of Saqqara. It is believed that this later monument was built by a successor of Zoser's named Huni, about whom virtually nothing is known. The step-pyramid builders were succeeded by the celebrated kings of the Fourth Dynasty, who built the true pyramids. These include the magnificent pyramids of Dashour and the world-famous triad at Giza. It is not inconceivable that Imhotep also planned these, even if he did not live to see them built.

THE MOUNTAINS OF THE STAR GODS 2

Here I am, O Ra (Sun-God), I am your son, I am a soul . . . a star of gold . . .

Pyramid Texts, lines 886–9

Modern archaeological scholars have cultivated a pristine ignorance of astronomical thought, some of them actually ignorant of the precession [of stars] itself.

G. de Santillana, *Hamlet's Mill*, p.66

I The Solar Theory

As we have noted, the true pyramid was prefigured in the step-pyramid of Zoser, the earliest large stone building yet discovered. It marked a turning-point in the development of Egyptian civilisation, which was approaching its zenith with the dramatic rise of the Fourth Dynasty. Zoser's monument required a huge leap in imagination as well as in technology and labour organisation, and it moved Egyptian funerary architecture in one giant leap from mud-brick mastabas towards the grandiose true pyramids.

The overwhelming consensus of Egyptologists is that the step-pyramid was a development of the mastaba but, unlike mastabas and the earlier burial pits, the step-pyramids were designed to be seen from afar and their outside look was as important as their internal burial chambers. The theory that the step-pyramids served as cosmic symbols is not revolutionary.[1] All over the world there are structures with a similar shape and meaning, from the *stupas* of south-east Asia to the stepped pyramids of central Mexico. They are invariably derived from the same basic archetype: the mountain or ladder from which the celestial world could be reached or which could serve as a platform for the sacerdotal duty of monarchs and rulers: a common concept in the sacred mythology of almost every nation. It was also part of the Egyptian heritage, for the hill of Annu was regarded in similar terms; it was the holy hill of Atum rising from the primal sea, and on its top stood his sacred pillar crowned with the Benben Stone, his sacred seed.

We do not know what originally stood on top of Zoser's step-pyramid but it may have been a replica of the Benben Stone,[2] which would be in keeping with the overall symbolism of the Pyramid Age. What we do know is that the later, true pyramids were capped with replicas of the Benben, for this was the name given to their crowning pyramidions. Examples can be seen today in the Cairo Museum,[3] which provide further evidence, if any were needed, that pyramid building was more than the creation of elaborate tombs.

The association of the Benben with Heliopolis, whose name means 'the city of the sun',[4] has led some Egyptologists to conclude that the pyramid shape is essentially a solar symbol, that it represents rays of the sun coming down to earth through the clouds. Thus the pyramid symbolises a crude stone ramp leading the pharaoh home to the sun. This is a relatively recent hypothesis and is an extension of the step-pyramid/ladder of the planets theory. It is repeated by Dr Edwards in his *Pyramids of Egypt* where he quotes Alexandre Moret: 'These great triangles forming the sides of the Pyramids seem to fall from the sky like the beams of the sun when its disk, though veiled by storm, pierces the clouds and lets down to earth a ladder of rays.'[5]

Commenting on this, Dr Edwards adds:

> When standing on the road to Saqqara and gazing westwards at the Pyramid plateau, it is possible to see the sun's rays striking downwards through a gap in the clouds at about the same angle as the slope of the Great Pyramid. The impression made on the mind by the scene is that the immaterial prototype and the material replica are here ranged side by side.[6]

This 'pyramid = the sun's rays' hypothesis has become deeply entrenched as a historical 'fact' and is quoted whenever the pyramids are discussed. However, this theory (and it is only a theory) unwittingly diverted the attention of researchers from the real symbolic purpose of these monuments. But we must first consider the history of true pyramid building, beginning with the work of Sneferu, first king of the Fourth Dynasty.

II The Sneferu Enigma

In the space of 500 years, from about 2700BC to *c.* 2200BC, more than thirty million tons of rock, enough to build Windsor Castle a hundred times over, was moved around the western desert near modern Cairo. It was used to construct pyramids,[7] some of which, like the Great Pyramid at Giza, are well over 140 metres high. They formed a huge royal cemetery for ancient Memphis, which we now call the Memphite Necropolis.[8] During this Pyramid Age, also known as the Old Kingdom, hordes of Egyptians worked like ants on this gigantic building site, while an equally large army of masons, goldsmiths, painters and scribes chipped, melted, brushed and scribbled away to prepare for the royal funerals.

To give some idea of the scale of these works, in 1980 an English-language newspaper in Saudi Arabi announced that a Franco-American consortium had signed a large contract to build the new University of Riyadh. The term 'large' hardly did justice to the sheer scale of the project: it was the largest single fixed-price contract awarded to a contractor in the history of constructional engineering and was valued at over US$1 billion. The logistics involved were staggering: 8000 workers on site, millions of cubic metres of rock and soil to move and hundreds of thousands of cubic metres of concrete to be poured. Even the site offices were on a monumental scale and an Olympic-sized swimming-pool and other leisure facilities would be provided for the staff.

Yet the Riyadh University project was modest in comparison with what happened at Dashour and Giza nearly 4500 years ago. When we compare the technology and resources available now – gigantic tower cranes, bulldozers and excavators, hydraulic cranes and so on – the unsatisfactory nature of the consensus view of Egyptologists concerning the pyramids becomes apparent. To refer to the gigantic pyramid complexes at Giza and Dashour as 'royal cemeteries' with 'royal tombs' is like calling the Palace of Versailles a house, or St Peter's in Rome a chapel. The building of the pyramids shows that the Pyramid Age was remarkable for great technological ability and daring innovation. But what exactly was the Pyramid Age? What sort of golden age was it?

According to Dr Edwards, 'the Pyramid Age, par excellence . . . covers the period beginning with the Third Dynasty and ending with the Sixth Dynasty.'[9] During this time some twenty-eight pyramid complexes were built along a stretch of desert running from Abu Ruwash in the north to Meidum in the south, the whole contained in an area about eighty kilometres long by some four kilometres wide. Yet such statistics are misleading, for they do not give us a balanced picture. It is often overlooked that most of the building took place in a very short period, the Fourth Dynasty, when seven of the twenty-eight pyramids were built; but such is the scale of these giant monuments that they account for more than 75 per cent of the total thirty million tons of material used during the entire Pyramid Age. Five of the seven, three at Giza and two at Dashour, have survived more or less intact to this day.

The first king of the Fourth Dynasty was Sneferu, father of Cheops. For reasons which Egyptologists have not yet determined, Sneferu and his architects abandoned the step-pyramid and launched the daring and huge smooth-sided pyramid design. Most scholars do agree that the motives for this dramatic change were religious, but what these motives were is not so evident. What is certain is that Sneferu's venture made the step-pyramid builders of the Third Dynasty look like village contractors. His builders raised not one but two gigantic pyramids, which, from present evidence, no other pharaoh ever attempted, before or after Sneferu. In addition to this massive building programme, Egyptologists believe that Sneferu's builders were able to satisfy him on another constructional enterprise; the upgrading of the step-pyramid at Meidum into a true pyramid by filling in the steps with masonry and adding the smooth casing-blocks.[10] But there is still much controversy surrounding this theory, and the Meidum pyramid, begun in the Third Dynasty and so far south of the Memphite Necropolis proper, cannot be treated in the same way as the other Fourth Dynasty pyramids.

To get a good idea of the engineering revolution which Sneferu initiated, compare the step-pyramid of Zoser, employing some 850,000 tons of material,[11] with Sneferu's two giants at Dashour, which together used nearly nine million tons. This amazing upsurge in engineering and organisational prowess has defied explanation, but it is obvious that something important inspired Sneferu, something which perhaps involved the thinking of the great master-builder-cum-priest – Imhotep. It is not simply the increase in scale but the fact that the technology was suddenly available to raise large blocks of stone, some weighing several tons, to a height of nearly 100 metres.[12] To make up the core of these pyramids, large blocks of hard limestone were quarried, transported, dressed, stacked then placed in perfect geometrical shapes which they have retained until today.[13] The question becomes not whether the pyramids were just tombs but what changed at the opening of Sneferu's reign that made it possible and indeed imperative to build pyramids on such a scale.

The textbooks do not provide a satisfactory answer to this question; they overlook the significance of the huge increase in activity during Sneferu's reign. Dr Jaromir Malek of the Griffith Institute at Oxford, in a recent book on the pyramids of Egypt, passes quickly over this subject, though he does say that 'the innovations introduced at the time [of the IVth Dynasty] were so wide-ranging that they must have had their origins in the sphere of religion rather than technology.'[14] Previously the renowned architect and Egyptologist, Dr Alexander Badawy, had merely written: 'At Meydum a true pyramid was obtained by filling up the steps of a layer pyramid . . . At Dashur Sneferu erected on a square base two true pyramids, one called the Rhomboidal (Bent) . . . It has been, however, observed that the upper part of this pyramid is poorly built . . . presumably finished in a hurry.'[15]

To describe the building activities at Dashour as being carried out 'in a hurry' is something of an understatement. If Sneferu's builders had only the same resources as their predecessors at their disposal – and there is no good reason to assume otherwise – they were in something more than a hurry. To make things even more difficult, the building activities were carried out on three sites: the two at Dashour which are two kilometres apart, and a third at Meidum, fifty kilometres further south, at least one day's journey up-river. Even if the work on the three sites was perfectly planned, there had to be a huge and complex organisation behind the scenes which would tax even today's large contractors. To cut, move and place nine million tons of limestone blocks in the space of perhaps two decades, in an epoch which did not know the wheel or the pulley and had no iron tools, is a factor which merits close scrutiny. Let us consider the scale of this achievement in the context of the Fourth Dynasty.

III The Golden Age of the Fourth Dynasty

What, therefore, could have happened *c.* 2650BC when Sneferu came to power and founded the great Fourth Dynasty? Dr Edwards was the first to bring some sense of proportion to the

Sneferu enigma, in 1947. Prior to his analysis, Egyptologists were faced with a perplexing situation, which involved the solitary Meidum pyramid in the extreme south of the Memphite Necropolis, originally built as a step-monument and later converted into a true, smooth-faced pyramid.[16]

There are no contemporary writings on the Meidum pyramid or elsewhere which give the name of its owner. However, in a temple nearby was found what Egyptologists technically call 'graffiti' (scribbles made by some passer-by). The graffiti were dated to the Eighteenth Dynasty – some 1200 years after the reign of Sneferu – and indicate that at that time (*c.* 1400BC) the Meidum pyramid was considered to belong to Sneferu.[17] Dr Edwards translated the graffiti, apparently written by a scribe called Aa-Kheper-Resenb, who lived during the reign of Tuthmoses III: '. . . in the forty-first year of the reign of Tuthmosis III . . . I came to see the beautiful temple of King Sneferu. . . .' Edwards also refers to graffiti from times earlier than this, as far back as the Fifth Dynasty (some 250 years after the reign of Sneferu), which mention his name in connection with Meidum.[18] This would normally be sufficient evidence to attribute the Meidum pyramid to Sneferu but there were the two other pyramids at Dashour to consider, with the 'southern . . . certainly built by Sneferu'[19] and evidence that strongly suggests he also built the northern one. An inscription found not far from Dashour dating from the reign of King Pepi I of the Sixth Dynasty, mentions the 'two pyramids of Sneferu'.[20] This was an official inscription, part of a royal decree exempting the priests of Sneferu from paying taxes, which must be regarded as sound evidence. Yet another, also found at Dashour and dating from the Fifth Dynasty, mentions the 'southern pyramid of Sneferu'.[21] Read together, these inscriptions imply that there were two pyramids, a southern and a northern, which belonged to Sneferu. The question then was, which pyramid should be considered the southern: the more southerly of the pair at Dashour, also called the bent pyramid because its angle of slope changes halfway up, or that at Meidum? Finally, on the tomb of a priest associated with the southern pyramid at Dashour, there is an inscription referring to the 'southern pyramid of Sneferu'.[22] This confirmed that the Dashour pyramid was the southern pyramid of Sneferu, and that its northern partner was also at Dashour. So where did the Meidum pyramid fit in? Egyptologists were at a loss.

Dr Edwards proposed a way out of this archaeological impasse. He pointed out that although ownership of two pyramids was unique to Sneferu, it could perhaps be understood to 'symbolise his sovereignty over Upper and Lower Egypt'; three pyramids, however, 'would seem to have no justification, practical or symbolic'.[23] It was a bold admission that archaeological evidence can be misleading. Edwards then proposed that until new evidence came to light, Egyptologists should consider Sneferu as the owner of the two Dashour pyramids, but that he had probably only converted the Meidum pyramid from a step-monument into a true pyramid.

It made good sense; the supposed builder of the original step-pyramid at Meidum was now said to have been the elusive Huni. Since Huni seemed to have reigned just before Sneferu, and would have been expected to have a pyramid of his own, this attribution meant that the Sneferu enigma could be regarded as sorted out and Egyptologists could get on with other pyramid problems.

But although Edwards's theory for Meidum made rational and even poetic sense, he issued the warning that since no inscriptions, either contemporary or of a later epoch, linked the mysterious Huni with the Meidum pyramid, it 'does not follow that his pyramid was indeed the step pyramid of Meydum . . .'[24]

This leads to the next problem: why did Sneferu build two pyramids when all other kings before and after him were satisfied with one? Did he intend to be buried in two places? Perhaps it is the words 'owned' and 'belonged' which are misleading. Could it be that the pyramids were not regarded as belonging to any particular king but rather to the royal lineage and the cult as a whole? Sneferu may have built two pyramids and converted a third, but perhaps they were not 'his' as we have hitherto thought. After all, medieval cathedrals, although built during the reigns of specific monarchs, do not belong to them even if they were interred in them.

If the ownership consensus of Egyptologists is to stand, how is it that Sneferu, who built two and perhaps even three pyramids, did not make it clear to posterity that he was the owner? There was plenty of space outside and in the Dashour pyramids for inscriptions to have been carved in capital letters. But none of the Fourth Dynasty kings put his name on the pyramid he supposedly owned. There is not one contemporary, official inscription, not even inside the Great Pyramid.

Ask yourself, if you had built the greatest tomb in history, after several decades of effort and cost, would you leave everyone guessing who had performed such a feat? It was not that the pyramid builders did not like official inscriptions on their monuments. From King Unas (last king of the Fifth Dynasty) onwards, pyramids had hundreds and hundreds of official inscriptions, leaving us in no doubt which kings built them.[25] Was it that Fourth Dynasty Egyptians could not write? Wrong again; many inscriptions exist in the vicinity of pyramids dated to the Fourth Dynasty and before. In the chapel of Queen Meresankh many hieroglyphic texts can still be seen. So this omission in the Fourth Dynasty pyramids is extremely odd and contrasts with the earlier mastabas and the inscribed pyramids of later dynasties.

Why did Sneferu, Khufu (Cheops) and the others not inscribe their pyramids? Never mind posterity, why leave the gods guessing who was responsible for these fine monuments? Did the Fourth Dynasty kings not regard themselves as individual owners of the pyramids? Is it possible that all the Fourth Dynasty pyramids were part of a single scheme, which required the building of seven different pyramids at specific locations?

The Fourth Dynasty as a whole is exceptional and stands out from the rest of the Pyramid Age. It seems to have arrived like the Egyptian phoenix who brings a new golden age, and over a short period of time carried out a magnificent programme of civil engineering, achieving a scale and standard of workmanship unparalleled until modern times.[26] Then, just as suddenly, it ended. The textbooks speak of 'religious upheavals' and 'civil wars',[27] but there is no evidence of these. If we are to find the answer we need to go back to the roots of pyramid research and question everything we have been told. Let us begin by looking again at the dating of the pyramids.

IV The Dating of the Pyramid Age

In the 1940s modern chronologists reshuffled the study of Ancient Egypt when they moved the early dynastic epoch forward by a millennium. Prior to this, the First Dynasty was thought to have begun earlier than today's estimates. In the 1830s Champollion, the father of scientific Egyptology and decipherer of the hieroglyphics, believed that the First Dynasty began *c.* 5867BC. Later the German Egyptologist, Karl Lepsius, moved this to 3892BC. Then Mariette, writing in the 1870s, reverted to 5004BC. Finally his colleague, Dr Brugsch, settled for 4400BC. Brugsch had apparently based his calculations on the simplistic assumption that there are, on average, three generations in each century,[28] but, for lack of anything better, his system was accepted for many decades by most Egyptologists.

Then in the 1940s the dating of the First Dynasty was again adjusted to *c.* 3100BC. This is constantly refined to 3150BC, 3300BC, 2900BC and so on, leaving us confused about what system we are to consider definitive. In any event, it should now be clear that the science of Egyptian chronology is far from perfect and relies on evidence dependent on personal interpretations and subjectivity. It is far from exact without tools that science has to offer, such as carbon dating and, in the case of symbolic architecture based on astronomical alignment as with the pyramids, the use of precession calculations.[29] Without them we wonder how early Egyptologists were able to establish such precise dates as '5004BC' and '5867BC'. Today chronologists use the prefix '*circa*' (*c.*) before a suspect date is given, to warn of a plus or minus variation. Yet we see such bold datings as the start of Sneferu's reign, given as *c.* 2686BC; *c.* 2584BC; *c.* 2614BC.[30] Not only are such figures misleading with their implications of accuracy, but evidence is rarely offered as to their computation. The earlier the epoch, the less accurate are such conventional dating systems, and concerning the Pyramid Age such estimates could easily be a century out, or more.[31]

In a letter we received from Dr Edwards during the preparation of this book, the date of *c.* 2600BC for Khufu's (Cheops's) reign was deemed 'most satisfactory'.[32] But the latest data, obtained in May 1993, from Rudolf Gantenbrink's laser measurements of the shafts in the Great Pyramid confirmed that even this 'most satisfactory' date may have to be adjusted by some 150 years to *c.* 2450BC,[33] to the time when those shafts were built.

When the pharaoh Sneferu came to power, the western desert near his capital at Memphis had already sprouted a few step-pyramids in the Memphite Necropolis. Today only that of Zoser remains. Much farther south was the lone step-pyramid of Meidum. Also in the Memphite Necropolis were many mastaba tombs, and farther north was the holy city of Heliopolis with its powerful priesthood. This was roughly the situation at the opening of the Fourth Dynasty.

Sneferu's bold decision to alter the traditional step-pyramid design resulted in the building of two gigantic true pyramids at Dashour. The credit should not necessarily go to Sneferu but, most likely, to his architect-priest who was either the legendary Imhotep or his successor. Imhotep is generally credited with the invention of stone masonry and the science of medicine. But Dr Edwards rightly pointed out that his title of Chief of the Observers suggests that he was an ancient astronomer who studied the motion of the stars.[34] This title was often assumed by the high priest of Heliopolis, which suggests that Imhotep, and those great masters like him, filled the function of high priest of that holy city.[35] Having so successfully completed the

Zoser step-pyramid complex, Imhotep, or his successor, was possibly fired by an even greater ambition: a unified master plan for the Memphite Necropolis which would allow full development of the royal rebirth cult.[36]

Sneferu is said to have died after reigning for nearly thirty-four years. With the coronation of his son, Khufu (Cheops), pyramid building was to reach its apotheosis.

V The Three Great Pyramids of Giza

Sneferu seems to have died peacefully *c.* 2480BC,[37] leaving his legacy of the two giant pyramids at Dashour. Some twenty-one kilometres north of Dashour was the elevated rocky plateau of Giza (the Mokattam Formation).[38] This plateau extended about 2.2 kilometres from north to south and about 1.1 kilometres across. It sloped gently from west to east, then dropped sharply near the contours of the Nile Valley.[39] It was on this imposing site that the eldest son of Sneferu, Khufu, launched the most ambitious engineering project in the history of building, which, with its two partners there, was to become the greatest wonder of the ancient world.

There is no satisfactory explanation why Khufu did not follow in his father's footsteps and build his pyramid at Dashour or indeed at Saqqara, where the ancestral step-pyramids and mastaba tombs were located. The easy answer is that the dominant position of the Giza plateau inspired him and his architect found the site ideal for a pyramid which would transcend those of Sneferu. But if this is true, why did Sneferu himself not select Giza instead of Dashour? Dashour is not much nearer to Memphis than Giza is and is a rather low site hardly visible from the palm groves which surround Memphis.[40] Indeed, uninformed visitors today are surprised to be told that there are two great pyramids other than those at Giza. So why did Khufu choose Giza, a site so far from his father's tomb? Perhaps he did not 'choose' Giza as such; perhaps it was already part of a master plan, intended to expand from south to north, devised during his father's reign? Georges Goyon, who was personal Egyptologist to King Farouk, was of the opinion that the Giza site was 'certainly chosen by the priest-astronomers because of certain religious and scientific factors', yet what these were Goyon does not tell us. We agree, however, that religion and astronomy motivated the ancient builders to move to Giza.

VI The Great Pyramid

Even today, in its ruined form and lacking almost all its glistening white casing-stones, the Great Pyramid is staggering. It broods over the surrounding desert and suburbs of modern Cairo, seeming more like some strange feature of the landscape than the work of human hands; indeed, more like a geometrical mountain than a building. The mathematician and journalist P. D. Ouspensky visited Giza for the first time in 1914, shortly before the outbreak of the First World War and wrote of the experience:

> The plateau is reached by a winding and ascending road which goes through a cutting in the rock. Having walked to the end of this road you find yourself on a level with the pyramids, before the so-called Pyramid of Cheops (Khufu), on the same side as the entrance into it. To the right in the distance is the second pyramid, and behind it, the third.
>
> Here, having ascended to the pyramids, you are in a different world, not in the world you were in ten minutes ago. There, fields, foliage, palms, were still about you. Here it is a different country, a different landscape, a kingdom of sand and stone. *This is the desert*. The transition is sharp and unexpected.
>
> . . . The incomprehensible past became the present and felt quite close to me, as if I could stretch out my arm into it, and our present disappeared and became strange, alien and distant.[41]

The Great Pyramid of Khufu, like the other pyramids, stands four-square, but it is in all its detail the most perfect. The first exhaustive survey of the monument in modern times was carried out by Sir Flinders Petrie in 1880–2. He used the latest equipment of the time and approached his task with great thoroughness. He found that the sides of the pyramid were indeed lined up almost exactly with the cardinal points of the compass: north, south, east and west. (The accuracy of this alignment is incredible, with an average discrepancy of only about three minutes of arc in any direction; this is a variation of less than 0.06 per cent.) He also measured the sides of the base as being 230.25 metres for the north side; 230.44 metres for the south, 230.38 for the east, and 230.35 for the west. Thus, although no side is identical to any other, the difference between the longest and shortest is only nineteen centimetres, less than 0.08 per cent of the average length.

Such degrees of accuracy, both in orientation towards the cardinal points and in keeping the base square and the sloping side perfect, are little short of miraculous when you consider the size of the structure. Its perimeter is almost one kilometre, with an area of over 53,000 square metres, enough to fit into it the cathedrals of Florence, Milan and St Peters, as well as Westminster Abbey and St Paul's.[42] It is indeed doubtful whether any of these later buildings exhibit the same accuracy as the Great Pyramid in their orientation or their structural execution. Although the pyramid contains several chambers, it is by no means a hollow building; it is mostly solid masonry and constructed from approximately 2.5 million limestone blocks. On average these weigh about 2.6 tons, to give a total mass of over 6.3 million tons.[43]

We can simply marvel at the craftsmanship and technological abilities of these ancient builders, for they not only orientated their monument towards the four cardinal points and kept the plan square and the slopes true, but they cased its four sloping faces with finely polished white limestone from the quarries at Tura on the other side of the Nile. Judging by the few facing stones remaining at the foot of the north side of the pyramid, these were even larger than those used in the core of the building and weighed some fifteen tons each. They were set so closely together that the blade of a knife could not fit between them. The casing-blocks were removed by the Arabs from the thirteenth century AD (some say to build the mosques of Cairo), but when intact the pyramid must have looked even more spectacular than it does today, glittering like a jewel in the sunlight.

It is now quite easy to clamber up and down the narrow corridors leading into the pyramids, for banisters are provided and there are wooden ramps with metal footings. The Giza pyramids are also electrically lit inside. Such luxuries were introduced in the 1940s, but exploration was not so easy for earlier travellers, as Ouspensky lamented in 1914:

> The floor is very slippery; there are no steps, but on the polished stone there are horizontal notches, worn smooth, into which it is possible to put one's feet sideways. Moreover, the floor is covered with fine sand and it is very difficult to keep oneself from sliding the whole way down. The Bedouin guide clambers down in front. In one hand he holds a lighted candle; the other he stretches out to you. You go down this sloping well in a bent attitude. The descent seems rather long – at last it ends.[44]

One thing that has not changed, of course, is the low height of the ceiling and the steepness of the gradient of this passage; it is only about 1.19 metres high and 1.04 metres wide and is sloped at 26 degrees 31 minutes 23 seconds to the horizontal. The passage plunges downwards, through the core of the pyramid and then through the bedrock that lies beneath it, for a total of 105.15 metres. It carries on horizontally for a further 8.83 metres before terminating in a roughly hewn chamber. The purpose of this chamber is unknown and the subject of much debate. It seems to be unfinished, and this has given rise to the so-called 'abandonment theory'. According to its proponents, this underground cavern was planned as the burial chamber of the king; for whatever reason and while the pyramid was still in its early stages of construction, this plan was abandoned in favour of building a new chamber high inside the pyramid itself (the Queen's Chamber). Later on this too was abandoned; a further corridor (the Grand Gallery) and a third room (the King's Chamber) being built. It is believed that the king was eventually buried in this last chamber, which contains a large sarcophagus, but no remains of the king's mummy or his funeral goods have been found and it is assumed that the pyramid was looted.

The abandonment theory has a certain attraction, but it runs against many practical requirements of building engineering. To have changed the design of the pyramid halfway through would have presented its engineers with nearly insurmountable problems. To have altered the design twice seems inconceivable, particularly when these alterations involved building the Grand Gallery, itself an extraordinary achievement, as well as the King's Chamber. We believe that the King's and Queen's Chambers, as well as the Grand Gallery which links them, were all part and parcel of the original design for the pyramid and were indeed essential features. There is no hard evidence to prove that the subterranean chamber was ever intended as the burial chamber of the king; indeed it may have been in existence before the pyramid was built as part of an earlier structure on the same site. We cannot be sure that this was so, but it is certainly facile to assume it was abandoned for technical reasons. The Egyptians were experts at building underground chambers and would not easily have been deterred from burying the king under the pyramid had that been his wish.

These days visitors are not allowed into the underground chamber; roughly eighteen metres from the entrance to the pyramid an ascending corridor begins. It is another back-breaking journey of approximately forty metres at a gradient of more than twenty-six degrees. As with the descending corridor, it runs exactly north-south (i.e., it is meridional). At the top of this corridor is the heart of the pyramid, the Grand Gallery, but before climbing this, a short journey along the horizontal corridor brings you into the Queen's Chamber.

As with so much of the Great Pyramid, the function of the Queen's Chamber remains a mystery. The academic consensus is that it was intended as the burial chamber of the king but, like the underground chamber, it was abandoned; it has recently been suggested that its entrance was too small to allow the granite coffer (now in the King's Chamber) to enter it.[45] This argument is not really tenable; a pharaoh capable of having such a massive and perfect pyramid built was unlikely to have altered his plans because someone had made his sarcophagus too large. Abandoning all this work for such a reason does not fit the probabilities.

The Queen's Chamber is not very large; only 5.74 metres from east to west and 5.23 metres north to south, its ceiling rising to an apex of 6.22 metres above floor level. In the east wall there is a niche, closely resembling a *mirab* (the prayer niche found in many mosques). The back of this niche has been cut away by treasure-hunters who no doubt hoped to find a secret chamber beyond. However, this is not the case; the niche was probably used originally to hold a statue of the king.[46] The walls are of carefully fitted, smooth limestone blocks. Though not as large or as elaborately finished as the King's Chamber, the Queen's Chamber looks no more abandoned than the rest of the pyramid. As it lies exactly on the pyramid's east–west axis, the Queen's Chamber must have been an important aspect, and it seems inconceivable that the Ancient Egyptians would have built it only to abandon it at the last moment.

The particular features of interest both to us and to Rudolf Gantenbrink are the two so-called 'air-shafts' in this chamber, which have for many years been seen as supporting the abandonment theory. These shafts, which have their counterparts in the King's Chamber above, were first discovered behind the walls of the chamber by a British engineer named Waynman Dixon in 1872.[47] As in the King's Chamber, one shaft is directed to the south and the other to the north. Further investigation soon revealed that, unlike those in the King's Chamber, these shafts do not run through to the outside of the pyramid, proving that they could never have functioned as ventilators as some have supposed.[48] In 1881 they were carefully investigated by Petrie, who measured their slopes and lengths with a clinometer. He concluded that they were not very long and that they seemed to serve no practical purpose. This was good ammunition for the abandonists, who concluded that the reason the shafts did not penetrate to the outside of the pyramid was that they were abandoned at the same time as the Queen's Chamber. The matter might have rested here had not Rudolf Gantenbrink proved that the shafts were much longer than hitherto assumed.[49]

Running from the level of the Queen's Chamber up to that of the King's Chamber is the amazing architectural creation known as the Grand Gallery. This is in many ways the most elaborate and mysterious feature of the whole internal system of the Great Pyramid, and words can scarcely do it justice. It is enormously impressive. It runs upwards at the same angle as the ascending corridor but instead of being a narrow, crouched tunnel it is 8.53 metres high. When you are inside, it gives the impression of being even higher as it sweeps towards the King's Chamber at the top end. It is a very curious structure indeed, for though it looks rather like a massive staircase, there are no steps as such. Yet it appears highly functional and was carefully finished in finely smoothed Tura limestone. Again Ouspensky provides a good description:

> In the construction of this upper corridor-staircase there is much that is difficult to understand and that at once strikes the eye. In examining it I very soon understood that this corridor is the key to the whole pyramid. From the place where I stood, it could be seen that the upper corridor was very high, and along its sides, like the banisters of a staircase, were broad stone parapets, descending to the ground, that is to the level where I stood. The floor of the corridor did not reach down to the ground, being cut short . . . at about a man's height from the floor. In order to get into the corridor from where I stood, one had to go up first by one of the side parapets and then drop down to the 'staircase' itself. I call this corridor a 'staircase' only because it ascends steeply. It has no

steps, only worn-down notches for the feet. Feeling that the floor behind you falls away, you begin to climb, holding on to one of the 'parapets'.[50]

Ascending the Grand Gallery is easier now, for there are short metal steps on either side leading from the level of the Queen's Chamber to the level of its floor. There are also handrails to help you up (and down) and a wooden ramp on the floor with metal anti-slip treads.

Although the Grand Gallery is now easier to explore, it is still overwhelmingly mysterious, especially when one realises that this curious room was ancient even in the days of Antony and Cleopatra. The walls are corbelled, so that the Grand Gallery becomes narrower towards its ceiling, and its cross-section design seems to echo the curious niche inside the Queen's Chamber, also corbelled. As with so much Egyptian architecture, it looks so ancient that it seems almost modern. There is a quasi-inhuman quality about the Grand Gallery that is hard to explain, as though it were not intended for people to walk up and down but to serve some other specialised or specific function. Many have remarked that the Grand Gallery looks like part of a machine, whose function is beyond us.

This is not a recent observation; the Neoplatonist Proclus drew attention to this in his fourth century commentary on Plato's *Timaeus*.[51] He claims that the Great Pyramid served as an astronomical observatory before it was completed, and that it was used as some sighting device for looking at the skies. This idea was taken up by a Victorian writer, Richard A. Proctor, who wrote *The Great Pyramid: Observatory, Tomb and Temple*, published in 1883.[52] He pointed out how the various corridors could have been used for observing the stars while the pyramid was being built; in particular, he suggested that the Grand Gallery could have been used to record the transits of stars. Proctor believed that the slots in the parapets were used to fix the position of a movable ramp used in this work. To understand this, it has to be remembered that because the Gallery is meridionally aligned to the southern sky, it could indeed have been used in this way at some time before the top part of the pyramid was built.

Then again, as some modern Egyptologists tell us, it may simply have been used for storing granite portcullis slabs. If so, the Egyptians went to enormous trouble to build such a store-room when a rough chamber would have sufficed. No one, however, has the answer to the riddle of the Grand Gallery and perhaps no one ever will.

Ascending the Grand Gallery brings one to the King's Chamber. Technologically this is the finest structure of all: it measures 10.46 metres from east to west and 5.23 metres north to south; its height is 5.81 metres. It can therefore be seen that while its floor area is exactly a double square of side 5.23 metres, it is just a little too high to be a double cube. Unlike the Queen's Chamber, which is lined with limestone, this is of smooth black granite brought from Aswan in Upper Egypt.[53] Whoever was responsible for building it was a master mason indeed. The granite blocks which make up the walls and ceiling weigh about thirty tons each and are perfectly smooth-faced. No mortar was used in jointing but, as with the casing-stones on the outside of the pyramid, the blocks were so perfectly cut and fitted that a knife blade will not fit between the joints. Fine jointing such as this would have been difficult with large limestone blocks; with huge granite blocks it is little short of incredible.

At the western end of the chamber is the mysterious granite sarcophagus. Although it is believed that this was the final resting-place of Khufu, there is not the slightest evidence of a corpse having been in that chamber, not a sign of embalming material or fragment of any artefact. No clue, however miniscule, has ever been found in this chamber or anywhere else in the Great Pyramid. This has led many to suppose that we have not yet found the true burial chamber of Khufu. Whatever the case, the sarcophagus in the King's Chamber has been badly damaged by souvenir hunters chipping pieces from its edges.

Finally, there are the two air-shafts of the King's Chamber to be considered. As in the Queen's Chamber, these rise from the north and south walls but they shoot right through the pyramid to emerge on its exterior. The four shafts found in the two chambers are all quite narrow, only some 20 × 20 centimetres in cross-section The belief that they were ventilators is a curious idea for a burial vault, and one not repeated in any of the other pyramids. As these shafts are central to our thesis, we will be returning to them later in greater detail. For the present, the consensus is that they were not intended to keep the pyramid ventilated, although one of the achievements of Rudolf Gantenbrink and his team was the fitting of ventilators in the shafts of the King's Chamber; this brought the humidity down from a stifling 90 per cent to 60 per cent, the same as outside. This is important when you consider the thousands of tourists passing through these chambers every day, each exhaling water vapour.

The completion of the Great Pyramid marked the high point of the Pyramid Age. When complete, it stood 147 metres (481 feet) high, some fifty metres higher than the larger of Sneferu's pyramids at Dashour, or equivalent to adding an extra fifteen-storey building on top. It also required that Khufu's workmen quarry and raise two million tons of stone more than the amount needed to build either of the Dashour pyramids. That Khufu was very serious about his pyramid is borne out by such textual information as we have about him.

There is, in the Berlin Museum, a document called the Westcar Papyrus. It dates from the New Kingdom but is undoubtedly a copy of a Fifth Dynasty original, for it tells the story of how this dynasty was ordained by the divine intervention of Ra, the sun god. The story takes place in the Fourth Dynasty during the reign of Khufu.

Wanting to be entertained, Khufu asks one of his sons, Djedef-Hor, to bring to him a magician called Djedi, an old wise man 'of one hundred and ten . . . who knows the number of the secret chambers of Thoth. Now His Majesty King Cheops (Khufu) spent all his time trying to find out the number of secret chambers of the sanctuary of Thoth so as to have the same for his own "horizon" . . .' He was therefore eager to meet the old magician. Horizon here means the Great Pyramid, for it bore the name 'the horizon of Khufu'.[54] Thoth, of course, was the ancient god of wisdom, depicted with an ibis head, who was reputed to have invented science and the system of hieroglyphic writing. His famous books, forty-two in number, were supposedly kept at Heliopolis and formed the basis of the state rebirth cult. In later times, Thoth was identified with the Greek god Hermes and was said to have been responsible for the planning and construction of the Great Pyramid.[55] When the magician Djedi arrives at court, Cheops asks him to perform some magical stunts and interrogates him: 'It is also said that you know the number of the secret chambers of the sanctuary of Thoth . . .'. To this Djedi replies: 'Please, I do not know their number, O king my Lord, but I know the place where it is . . . there is a chest made of flint in the building called "the inventory" in Heliopolis. It is in this chest.' Djedi then says that he cannot get it and neither can the king; only three as yet unborn kings in the womb of a priestess of Heliopolis will have that privilege. These are the first three kings of the Fifth Dynasty: Userkaf, Sahura and Neferirkara.

Unfortunately, the Westcar Papyrus does not tell us what happened to the chest in Heliopolis or whether Khufu did obtain it and use the information it contained in the building of his pyramid. We are left wondering whether he did discover a secret chamber of Thoth at Heliopolis and, as is hinted in the papyrus, build a secret chamber of his own inside the Great Pyramid.

Work went on at Giza long after the death of Khufu. He was succeeded by Khafra (Chephren): who built another giant pyramid next to the Great Pyramid. Though only a few metres short of the first, the second appears taller because it stands on a slightly higher part of the Giza plateau. After Khafra came Menkaura (Mycerinos), who built a smaller pyramid, 65.5 metres in height. By any other standards, the third pyramid is a giant, but it is dwarfed by its neighbours on the Giza plateau.

Six kilometres north-west of Giza is Abu Ruwash, where

a son of Khufu, King Djedefra, built his pyramid, but this has not survived time and plunder. It is now a pitiful heap of rubble and hardly recognisable as a pyramid. Its dimensions are not known for sure, but it seems to have been a large structure, perhaps similar in size to that of Menkaura at Giza. Another obscure pharaoh named Nebka, perhaps a brother or son of Khufu, planned a pyramid at Zawyat Al Aryan, a site about five kilometres south-east of Giza. It was either never finished or was dismantled in later epochs and used as a ready-made quarry.[56] With Nebka the Fourth Dynasty came to an end. What happened next is unknown to Egyptologists; we are faced with an apparent loss of will and consequent decline in pyramid building after the Fourth Dynasty.

To put their achievements into context, it may help to compare the sizes of their known pyramids. The table gives the approximate size and mass for each.

HEIGHTS AND MASS OF FOURTH DYNASTY PYRAMIDS

Location	Height, metres	Mass, million tons
Dashour South	102	3.59
Dashour North	101	4.00
Giza (Khufu)	147	6.18
Giza (Khafra)	140	5.28
Giza (Menkaura)	65	0.57
Abu Ruwash	unknown	0.50
Zawyat Al Aryan	unknown	1.50
Total		21.62

To this twenty-one million tons must be added the mass of rock to raise boundary walls, temples, causeways and other structures forming part of a pyramid complex. We can conservatively add a further one million tons of limestone and granite, and this twenty-two million tons[57] represents more than 80 per cent of the rock used during the whole Pyramid Age. The Fourth Dynasty, literally, towers above those which preceded and followed it.

VII The Collapse of the Fourth Dynasty

Jaromir Malek, director of the Griffith Institute of the Ashmolean Museum, has claimed that we do not need knowledge of architecture or history to know which pyramid came first:

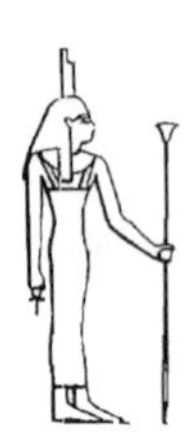

> It is enough to look at their present silhouettes: the step pyramid . . . is of the Third Dynasty . . . the pyramids proper which present a clean and sharp outline against the sky date from the Fourth Dynasty; those of the Fifth and Sixth Dynasties are now ragged shapes resembling huge piles of stone block and rubble . . .[58]

It is obvious to anyone visiting the pyramids that after the Fourth Dynasty there was a sharp decline in the skill of pyramid building. The kings of the Fifth Dynasty built five small pyramids at Abusir, about nine kilometres south-east of Giza, and a further two small pyramids at Saqqara, not far from Zoser's step-pyramid. All of these were rather poorly constructed, and the workmanship of the inner core, which has mostly collapsed, is very much shoddier than that of their illustrious predecessors of the Fourth Dynasty. All the Fifth Dynasty pyramids are now mere heaps of rubble, some more like mounds than pyramids.[59] Four small pyramids were built by the Sixth Dynasty pharaohs at Saqqara, all about fifty-three metres high and of even shoddier workmanship. With these last 'the Pyramid Age par excellence', as Edwards puts it, came to a close.[60]

The Fifth and Sixth Dynasty pyramids required some 2.75 million tons of limestone for their construction, less than half the mass of Khufu's pyramid at Giza. This, and the obviously shoddy workmanship involved, implies that something drastic must have happened at the close of the Fourth Dynasty, something as inexplicable as the sudden emergence of the Fourth Dynasty with the rise of Sneferu and his ambitious project at Dashour.

The Fifth and Sixth Dynasty builders had the experience of the great Fourth Dynasty to fall back on, so from an engineering point of view we would have expected a progression and not a regression in the skills of raising monumental pyramids. Some Egyptologists believe the problem was one of social upheaval or economics. But if the later dynasties could not match the Fourth in the scale of the projects they undertook, at least they should have been able to sustain the quality of workmanship.

It is almost as though Egypt experienced a technological exodus at the end of the Fourth Dynasty, a brain and skill drain that depleted the pharaonic state. During the Fourth Dynasty, the Egyptians were supreme master builders, then suddenly, within a generation or so of their demise, there was an amazing loss of skill in the art of pyramid building. This is so pronounced that even the most conservative of architectural Egyptologists, Dr Alexander Badawy, describes the Abusir pyramids as being 'strikingly poorer than the megalithic Fourth Dynasty structures'.[61] Visitors to the Abusir site are hard pressed to believe that such pitiful heaps were once geometrical pyramids.

Egyptologists still debate the events that led to what they call the collapse between the end of the Fourth and the start of the Fifth Dynasty. They talk of socio-political upheaval, but Dr Malek claims that 'the Old Kingdom was not brought to its knees by an upheaval caused by a popular uprising . . . no large-scale invasion from abroad took place. . . .' He believes that there occurred a weakening of the state's authority caused by a 'gradual shift in the ownership of land from the central authority to cult and temple establishments and the nobility as a whole'.[62] Yet, as far as we know, there is no evidence to confirm this; there are no land deeds or decrees to support such contentions. Edwards, on the other hand, feels that there was a violent cultural or religious change which caused a shift in authority to the priests of Ra, the sun god, whose centre was at Heliopolis. But he, too, admits that 'documentary records are lacking' to support this theory.[63] If the truth is told, nobody knows what happened; conventional reasoning cannot explain the evidence we have before us. All we can say is that whatever happened at the end of the Fourth Dynasty caused the eventual collapse, as Malek describes it, of the great Pyramid Age.

The Giza pyramids are the crowning achievement of Ancient Egypt and the ancient world. It also seems that the dynamic momentum set by the Fourth Dynasty was slow in fading; although Fifth and Sixth Dynasty pyramids were smaller and shoddier, the urge to build them was still there, and we get the impression that it was not a collapse that occurred but something more like the handing over of the state's authority to a less experienced government after a large-scale event.

VIII Evidence of a Master Plan

In 1934, towards the end of the Great Depression, a successful American architect, James A. Kane, visited Dr John Wilson, then director of the Oriental Institute at the University of Chicago. Kane had brought a large folder containing detailed drawings, calculations and a geo-survey analysis, not for a new office block or some mansion in New England, but of the Giza plateau and the three great pyramids which stand on it. Wilson's first reaction was to try to persuade Kane to drop this hopeless business of 'solving the mystery of the pyramids' but then, in his own words: 'I found myself constantly falling back to the cry of "coincidence"! Now coincidence may be invoked once or even twice, but when several divergent elements coincide and coincide again, coincidence becomes conformity rather than chance.'[64]

What the architect was showing him seemed obvious: the Giza pyramids, seen as a whole, were built in accordance with an architectural master plan. In his thesis, 'The Ancient Building Science', Kane was presenting a detailed analysis of the geo-architectural aspects of the Giza pyramids which showed conclusively that each of the three great pyramids was part of a single, unified plan, one which must have been devised from the outset of the great enterprise at Giza. We do not propose to go into the details of his analysis, but Kane saw that the three Giza pyramids were developed from a plan based upon geometrical and surveying principles which he believed were related to astronomical observations. Even in the 1930s, most Egyptologists were aware that the pyramids were set out and orientated using astronomical observations. For example, the bases of the pyramids are all set along meridians, so that each side of their square bases faces one of the four cardinal points. That the entrances to the pyramids are virtually all on their north faces, and that their internal systems are designed to run along their north to south axes, indicates that this meridional setting was paramount.

Recently the American archaeologist, Martin Isler, reiterated this fact in connection with Khufu's pyramid, saying 'accurate orientation could only have been achieved by using celestial bodies'.[65] The accuracy is indeed stunning; there is an average deviation of only 1.8 arc minutes, minimal for such a large monument.[66] The celestial body or bodies which served as the orientation target could not have been the large discs of the sun or moon (as Isler supposed) but must have been a pinpoint of light, which strongly implies a star. Edwards adds support to this stellar hypothesis with the opinion that 'it seems more likely that the high degree of accuracy was achieved by astral rather than by solar observations'.[67] This becomes obvious when we know that the Ancient Egyptians were avid stargazers. The priests watched the night sky not only for religious reasons but for telling time by the rising of stars and their culmination as markers to some natural star-clock mechanism based on the apparent daily and annual motion of the stars. R. O. Faulkner, the 'definitive' translator of the Pyramid Texts, also writes that 'it is well known that the Ancient Egyptians took great interest in the stars, not only for practical purposes . . . but inscribing star-maps and tables in their coffins and tombs . . . in which the stars were regarded as gods or as the souls of the blessed dead.'[68] Indeed it has often been demonstrated, not least by Dr Edwards, that the meridian line to set a pyramid's square base could best be achieved by observing the stars. Everything points to astral methods having been used, not only because of the accuracy this gives but because we know that the Ancient Egyptians had, at the outset of the Pyramid Age, a strong stellar religion deriving from an ancestral cult.[69]

All this might seem obvious to us, but James Kane's ideas made little headway in Egyptology circles. Although he published his thesis, it was put to one side and forgotten. Several decades later, in 1984, the American Research Centre in Egypt (ARCE) launched the Giza Plateau Mapping Project. This was to be carried out during two seasons in the period 1984 to 1986. The team leader was Mark Lehner, an American-born Egyptologist from Yale University. Two major reports were published in ARCE newsletters in Egypt before Lehner published his full report in a prestigious German Egyptological journal.

Lehner's reports are largely based on surveying and geological data. Curiously and in view of his earlier literary work,[70] he was not much concerned with the cultic aspects of whatever plan might exist, nor indeed with the symbolic architectural and astronomical messages the monuments might contain; his focus was on the geomorphy of the site and the need to determine exact co-ordinates for the analysis of geological formations in the Giza plateau. While many waited for new physical evidence for what some engineers already suspected – that the Giza pyramids were part of a unified master plan – all that came out of the 1984–6 surveys was a mass of complex geological and surveying data which raised more questions than it answered. Although Dr Lehner had performed an excellent geological and land-survey exercise, the burning questions related to a unified plan were not answered but further obfuscated by technical jargon. Yet he was awe-struck by the grand scale of the pyramids of Giza and Dashour which the Fourth Dynasty raised, and wrote that 'when graphed against time, this brief period of the most monumental architecture stands out as a sharp peak dwarfing the material invested for royal construction prior or subsequent to the reigns of these kings'.[71] Lehner later reported that an obvious diagonal alignment existed running just east and close to the monuments. This line projects from the south-east corner of the first pyramid (Khufu) to the south-east corner of the third pyramid (Menkaura), later referred to as the 'Lehner line' by other researchers.[72]

Kane and now Lehner were thus pioneers of a new avenue of research and people began to think about a unified ground plan for Giza. They were not the only ones to think along these lines; at least two other researchers pursued it further and their results were more extraordinary still.

IX A Unified Ground Plan

As is often the case with valid theories which evolve from the convergence of diverse data, the 'master plan theory' had popped up even before Lehner's survey was finished. A similar suggestion had come from John Legon, a self-employed physicist living in Surrey, England. He first expounded its basis in the *Reports of the Archaeology Society of Staten Island*.[73] In 1988, and in greater detail, he wrote a paper entitled 'A Ground Plan at Giza' and this was published in the Oxford journal *Discussions In Egyptology*.[74] Legon investigated the 'possibility of a positional relationship between the three pyramids' at Giza.

His thesis was passed to me late in 1988 by Dr Edwards, who seemed interested in Legon's theory of a unified plan at Giza, which was as follows:

> The placing of the three pyramids in a single ground plan was obviously an ambitious project, and one which indicates that the architects and builders of the Fourth Dynasty had a much greater control . . . than had hitherto been recognised. They were apparently able to dictate, for example, the small dimensions of the Third Pyramid, despite the presumed desire of

> Menkaura to have a monument equal to those of his predecessors. Since the three large pyramids of Meydum and Dashur appear all to have been built by Sneferu, it seems possible that at the outset, Khufu himself might have aspired to the construction of the three pyramids of Giza in a single unified ground plan.[75]

Legon showed mathematically that the three Giza pyramids fitted inside a rectangular perimeter having the north-south side as 1732 cubits[76] and the east-west side as 1414 cubits. It occurred to him that a basic modular unit of 1000 cubits was used, and could be expressed as $1000\sqrt{3}$ and $1000\sqrt{2}$. Since these sides were of a right-angled triangle, the diagonal could be expressed as $1000\sqrt{5}$. He concluded that such geometrical and mathematical harmony could not be the product of coincidence. The notion of a master plan at Giza was now getting strong support from other quarters, but Legon, intent on proving that there was evidence of a master plan, also omitted to investigate the religious or cultic motives for it.[77] The question that still hung in the air was, and still is, what does the master plan express?

In February 1988 a teacher and geologist, Robin J. Cook, published a paper entitled 'The Giza Pyramid: A Design Study'.[78] Cook expanded on the findings of Lehner and Legon and added some ideas of his own to show that 'the Giza pyramids were designed according to a system of geometrical ideas, and that the site was planned as a whole . . .' Cook pointed out that a geometrical axial system could be shown to link the central pyramid, that of Khafra, with the small satellite pyramids next to the first and third pyramids. The main angles exposed were 60 degrees and 26.5 degrees; 60 degrees is the angle of the isosceles triangle and 26.5 degrees is produced by the diagonal of the double-square. This angle of 26.5 degrees could also be found in the main passageways of the Great Pyramid and the double-square of the floor of the King's Chamber; again defying the limits of coincidence. Cook, unlike Legon and certainly unlike Lehner earlier, sensed a powerful symbolism behind the plan, evidence which revealed the use of geometrical and geo-architectural patterns to express an ancient system of numerical philosophy. He rightly observed that:

> The Giza Pyramids represent a symbolic statement written in stone and the language of a mathematical philosophy. The Giza group probably represents *a symbolic expression of the Heliopolitan myth* [my emphasis] . . .[79]

Yet Cook seemed unable to say what the symbolic statement was that was written in stone.[80] He and Legon had demonstrated the advanced state of Egyptian geometry at the time the pyramids were built and that it could be applied in practical ways, but this was not enough to explain the Giza pyramids or their layout.

It seems we need to look further for these answers, not at Giza but in the small Fifth and Sixth Dynasty pyramids at Saqqara. There are inscribed, inside the little pyramid of Unas, some extraordinary texts.

3 THE DISCOVERY OF THE PYRAMID TEXTS

The Pyramid Texts . . . constitute the oldest corpus of Egyptian religious funerary literature now extant. Furthermore, they are the least corrupt of all such collections of funerary texts, and are of fundamental importance to the student of Egyptian religion . . .
– R. O. Faulkner, *The Ancient Egyptian Pyramid Texts*

Alexander Piankoff, a translator of the Pyramid Texts . . . was seriously opposed to the present trend of using the religious texts primarily for the search of dates and the accumulation of separate facts . . . [he] aimed at letting the writings speak for themselves and thus evoke the symbols and prototypes of religious thoughts . . . the Pyramid Texts were aimed at insuring the same rebirth for the dead king as that of the god Osiris-Orion . . .
– Jane B. Sellers, *The Death of Gods in Ancient Egypt*

I The Day of the Jackal

Hidden inside some of the Fifth and Sixth Dynasty pyramids are the oldest religious writings yet discovered in the world. These, for obvious reasons, are known as the 'Pyramid Texts'. Given their extraordinary antiquity, it seems strange that they are not better known to the public. Most people have heard of the Dead Sea Scrolls, which are from a much later epoch (*c.* 100BC) and much less interesting documents. It is curious that the Pyramid Texts have been so neglected by most people, a mystery in itself.

When I first came across them in 1979 I was astounded and wondered why hadn't I heard of them before. Talking with friends in Cairo, I discovered that many Egyptians were ignorant of their existence. I was convinced that they were far more important than we had been led to believe, and decided to examine them more closely. I soon perceived that everything about these ancient texts is mysterious, even their discovery, which happened in a most curious way.

During the winter of 1879 a rumour was circulating in Cairo that ancient inscriptions might exist inside the small and unexplored pyramids at Saqqara. The rumour gathered momentum and aroused the usual mixture of scepticism and excitement until it reached the ears of Professor Gaston Maspero. He had recently arrived in Cairo to take charge of the Mission d'Archéologie Française and was eager to further his career in Egypt. An experienced archaeologist and brilliant philologist, Maspero knew only too well that the biggest archaeological finds often begin with just such a rumour, a whisper in the markets, and this one had a feel of truth. It seemed to confirm what he had secretly suspected about the otherwise silent pyramids of Egypt. He decided to investigate.

Apparently, a jackal or desert fox had been spotted at dawn immobile near a crumbled pyramid in the necropolis of Saqqara. It was as if the animal were taunting his lone human observer, a *reis* or head workman, and was almost inviting the puzzled man to chase him. Slowly the jackal sauntered towards the north face of the pyramid, stopping for a moment before disappearing into a hole. The bemused Arab decided to follow his lead. After slipping through the narrow hole, he found himself crawling into the dark bowels of the pyramid. Soon he emerged into a chamber and, lifting his light, saw that the walls were covered from top to bottom with hieroglyphic inscriptions. These were carved with exquisite craftsmanship into the solid limestone and painted over with turquoise and gold. The *reis* had stumbled across one of the greatest archaeological discoveries of the late nineteenth century and the coded messages which eventually led to the resolutions of the mystery of the pyramids.

There is a certain irony that the discovery was made by following a jackal. In Ancient Egypt there were two jackal gods, though they were probabaly different aspects of the same divine archetype. The first and best known was Anubis, who in Egyptian funerary paintings is always shown supervising the ritual 'weighing of the heart', the dreaded final reckoning of the dead that decided whether or not a soul could enter into the court of Osiris. Wooden sculptures of Anubis were also made and placed as guardians inside the tombs of pharaohs; a beautiful example of one of these (now in the Cairo Museum) being the ever-watchful guardian found in the tomb of the boy-king Tutankhamun. The other jackal was Wepwawet or Upuaut, the 'opener of the ways'. It was after him, of course, that the German team named their famous robot.

The distinction between Anubis and Upuaut is not clear from the ancient texts, but, as Robert Temple pointed out in *The Sirius Mystery*, Anubis was seen as linked with Sirius, the brightest star in the constellation Canis Major (the Great Dog). Upuaut seems to have been connected with the northern constellation which we now call Ursa Minor. (The jackal is also involved with our quest for solution of the Orion Mystery and I was to encounter 'my' jackal at Giza just before making an important discovery.)

II Parlez-vous Français?

The discovery of the Pyramid Texts is shrouded in controversy. The late 1870s were, admittedly, a confusing time in Egypt. The mood was of imminent civil disturbance and even of civil war, and there were signs of revolt against foreigners and the puppet khedive, Tewfik Pasha.[1] A military fleet was preparing to sail from Britain to intimidate the rebels and their leader, Ahmed Arabi, who had been threatening the khedive's authority and harassing and murdering Europeans in Cairo and Alexandria.[2] Amid the political instability the rumour of the jackal's find added worry and confusion for the foreign archaeologists in Cairo, who were concerned with safeguarding their livelihoods as well as archaeological treasures.

The credit for the discovery of the Pyramid Texts is generally given to Gaston Maspero, but the true sequence of events that led to the discovery are far from clear. It is well documented that he was the first to enter the pyramid of Unas on 28 February 1881, but there can be no doubt that two other text-bearing pyramids had already been secretly explored by Auguste Mariette (1821–81), director of the Egyptian Antiquities Service.[3]

The story goes that the Arab *reis*, probably disappointed at not finding any 'real' treasure inside the small pyramid, reported his find to the authorities responsible for antiquities, which meant Auguste Mariette, the most senior Egyptologist of his day, who had donned the title of pasha. Mariette was a native of Boulogne, and had been in Egypt since 1851. He had become famous a few months after his arrival in Egypt, when he had discovered the Serapeum at Saqqara, a huge labyrinth of underground galleries containing dozens of the massive sarcophagi of the sacred Apis Bulls of Memphis. This made him a good friend of Khedive Said and later of his son, Ismail, which gave him considerable influence in Egypt. Mariette founded the Services des Antiquités, the prototype for the Egyptian Antiquities Organisation, and the Boulag Museum, which eventually became the Cairo Museum and moved to its present location in Tahrir Square. Mariette became the first director of the Services, then a position of power in Egypt, since it controlled the trade of antiquities and the concessions to foreign bodies wishing to excavate.

By 1880, when the Pyramid Texts were discovered, Mariette had become a household name and his reputation as an archaeologist was immense. He had also a reputation for stubbornness and authoritarianism which, on more than one occasion, caused political trouble with his mentors.[4] His star was setting; he was tired and sick, and had lost his wife and a child in an outbreak of the plague in Egypt. Mariette brooded over what he regarded as his private empire, the Memphite Necropolis where, among other treasures, he had discovered the Serapeum.

It was well known to all that Mariette had been something of a rebel in his youth. He had originally been sent to Egypt by the Louvre, not to excavate but to look for Coptic manuscripts, which he was given funds to purchase. Instead, relying on his intuition, Mariette used the money to carry out unauthorised excavations at Saqqara. Luckily, his hunch proved right, and he discovered the Serapeum. The Louvre's curators forgave him and sent him more money to carry on with excavation work.[5] However, this was all bygone days; now, as an old and tired man, he refused to allow his younger colleagues the freedom he had once enjoyed. When the rumours concerning the pyramid were brought to his attention, he refused to follow them up or let anyone else do so. In spite of entreaties by Maspero and others, Mariette maintained a rigid and patronising stance, claiming that it would be a waste of time and money to enter these unexplored pyramids. His argument was that as pyramids were tombs they could not 'speak'; they were obviously *muettes* (mute), and he insisted that they could not possibly contain inscriptions. His colleagues, including Maspero, decided it was best to let matters stand.

On the face of it, Mariette seemed to have a valid point. It had to be admitted, even by the optimistic Maspero, that all the pyramids opened so far, including the great pyramids of Giza, contained no contemporary inscriptions whatsoever. The only writings found inside were graffiti of no great value.[6] There was no reason to believe that the smaller pyramids at Saqqara would be different. There was only the jackal rumour and, though Maspero took it seriously, Mariette was not impressed and reiterated his objection by asking 'If the pyramids contained texts, they would not be just tombs, would they?'[7] Maspero said later: 'One knew quite well the opinion of Mariette on this subject of pyramids: in the preface to his unfinished work on the mastabas, he wanted forcefully to prove not only that they contained no texts, but that they never had contained any inscriptions and that it would be a waste of time and money to want to open them . . .'.[8]

Early in 1880, however, the money problem at least had been solved. The French government made a generous donation to the Antiquities Service of 10,000 francs, on the understanding that at least one of the unopened pyramids at Saqqara should be explored. Maspero had been urging that the funds be sent in the hope of softening Mariette's opposition. It worked, but not in the way Maspero had hoped:

> The work, started in April 1880 under the guidance of the Reis Mohamad Chahin, resulted in the discovery of two ruined chambers and a corridor, covered with hieroglyphs. The imprints of the inscriptions, carried out by Mr Emile Brugsch-Bey, were handed to me by Monsieur Mariette, without indication of their origins, asking me to examine them and translate them. A first glance made me recognise texts which came from the pyramid of Pepi I.[9]

Maspero claimed that Mariette insisted these texts were not from a royal pyramid but from the large mastaba tomb of a nobleman:

> Monsieur Mariette was so biased in favour of his theory of 'dumb' pyramids, that he at first did not want to admit that the tomb the inscriptions had come from was a pyramid, and that it had entombed Pepi I: according to him they had only found a mastaba of large size belonging to a common individual . . .[10]

At last, on 4 January 1881 Mariette relented. This, after all, could be his last chance to be privy to the secrets of the pyramids. Reluctantly, he gave instructions to his German assistant, Emile Brugsch, to investigate this irksome 'jackal rumour'.

A few days later Brugsch reported to Mariette that the *reis*'s story had been correct. It was in a pyramid and not a mastaba that the inscriptions had been found.[11] But by now the great archaeologist was on his deathbed and, ironically, never saw the texts. On 19 January 1881 Mariette died at Boulag, near the famous museum he had created, and his embalmed body now rests inside a sarcophagus in the courtyard of the Cairo Museum of Egyptian Antiquities. A bronze statue of Mariette dominates the scene, with a plaque, 'A Mariette Pasha, L'Egypte Reconnaissante'.

Maspero was Mariette's obvious successor, and was immediately appointed Director of the Services des Antiquités. It was clear to everyone what his first move would be: with the official authority his new position brought, the full exploration of the neglected small pyramids in the Memphite Necropolis was on a secure footing.

Thus it was that in the second week of February 1881, under a glorious winter sun, Maspero embarked upon the operation with quasi-military zeal. He decided to 'attack along the whole front of the Memphite Necropolis, that is from Abu Roash [Ruwash] to Lisht . . .'[12] The pyramids of Pepi I and Merenre had already been opened by Brugsch and now 'rapid success was to follow. Unas was opened on the 28 February, Pepi II and Neferirkara on 13 April, and Teti on 29 May . . .' Excavations went on until late in 1882 on other pyramids with no further inscriptions found, but Maspero was proud to report that 'in less than a year, five of the so-called "dumb" pyramids of Saqqara had spoken . . .'[13]

This was more than he had ever dreamt would be found; literally thousands of lines of hieroglyphs had now been discovered. One can feel Maspero's excitement as he explains the quantity of writings involved. 'The result', he wrote, 'is considerable. The inscribed pyramids at Sakkara have given us almost 4000 lines of hymns and formulae, of which the greater part were written originally during the prehistoric period of Egyptian history.'

His conclusion as to the date of their original composition, even by conservative estimates, brings us to a period around 3200BC, which is almost two millennia before the compilation of the Old Testament and over 3400 years before the first Christian gospels were written. The Pyramid Texts are certainly the oldest religious corpus of writings discovered anywhere in the world.

Of the five pyramids involved, the one which was to yield the greatest number of texts was that of Unas, last of the Fifth Dynasty kings (*c.* 2300BC). The pristine texts in this pyramid were not only the finest in the collection but the oldest. Maspero was the first person to enter the chambers of Unas and see the texts. He had to crouch as he made his way through the low, descending passage until he reached the sarcophagus chamber with its wonderful pitched ceiling. Here he (like Adrian and I a century later) gazed with awe at the wonderfully cut hieroglyphs inscribed on the walls.

Maspero now had the difficult task of translating and interpreting what he had discovered. He wrote, 'The texts which cover [the walls] are of three kinds: ritualistic texts, prayers and magical formulae.'[14] It was an unfortunate choice of words, for comments like this were to undermine the significance of the find. This was one of the most exciting archaeological discoveries ever, but by labelling the Pyramid Texts little more than *grimoires* of pagan superstition, he made them seem inconsequential. Maspero failed, like many others after him, to detect the astronomical content of the writings and the expression of a potent esoteric wisdom.

It took the best part of five days for Maspero, with the help of Emile Brugsch, to copy down the texts from Unas's pyramid; within a few weeks he had a rough translation ready for publication in the official journal of the Mission Archéologique d'Egypte. He wrote later:

> I do not hide the fact that this tentative translation was rather rash, and I perhaps should have waited longer; I none the less thought that Egyptologists would be more grateful to me for a quick publication rather than waiting for an in-depth study, and would therefore forgive me the errors in interpretation in favour of the importance of the texts.[15]

Maspero's confession proved necessary, because he was precipitate in his interpretation of the Pyramid Texts. Unfortunately a great deal of misunderstanding about them was caused not only by him but by other Egyptologists in the early part of the twentieth century. In their enthusiasm to bring out translations and commentaries, they depended as much upon gut instinct as anything else and this tended to be loaded with Christian bias.

The greatest culprit was an American Egyptologist named James Henry Breasted, who made a serious attempt at interpreting the texts in 1912. Breasted was to see in the Texts something that was not there at all: the remnants of a solar cult versus stellar cult rivalry, with the stellar cult in decline and there only for nostalgic reasons. He was thus to write:

> stellar notions have doubtless descended from a more ancient day when the stellar notion was independent of the solar . . . it is evident that the stellar notion has been absorbed by the solar . . . the solar beliefs predominate so strongly that the Pyramid Texts as a whole and in the form in which they have reached us may be said to be of solar origins.[16]

Breasted concluded that the stellar cult deserved little attention; all his attention went to what he saw as the principal theme of the Pyramid Texts, a solar cult. The inevitable result was that the pyramids were allocated a solar pedigree by Breasted; such a conclusion put a solar stamp on them and their symbolic purposes that was going to be very hard to shift, for Breasted was no ordinary Egyptologist. By the end of his 'brilliant career' his list of credentials and titles filled two pages, and he was dubbed 'the real founder of Egyptology in the New World'.[17]

Breasted (1865–1935) came from 'sedate Mid-western stock and had once planned to prepare himself for the ministry'; his interest in ancient peoples eventually drew him to the study of 'Bible lands', although he always 'retained a strong sense of mission'.[18] He began his working life as a clerk in local drug stores, graduating in pharmacy in 1882. He then went on to study Hebrew in Chicago and moved to Yale University in 1890–1. There he was drawn to the study of Egyptology, which remained his life-long passion. In 1892 he went to Berlin and studied under the German philologist, Dr Adolf Erman. He gradually made a name for himself and attracted the attention and friendship of J. D. Rockefeller Jr., who, in 1924, gave him a grant which Breasted used in part to found the Oriental Institute at Chicago, America's first Egyptological seat. Further gifts from Rockefeller allowed Breasted to turn the Oriental Institute into the leading Egyptological institute of the New World, commanding the deep respect of scholars and students alike.[19] With this status and academic authority, few would have dared to challenge his established views.

There is no doubt that Breasted's contribution to Egyptology is immense, but this does not alter the fact that his biblical bias and his personal vision of a monotheistic solar religion which he sought to graft on to the Pyramid Tests nearly closed the door to a fresh interpretation of them. There were many who sensed that something was adrift in his interpretations, and that the astronomical and stellar aspects of the Texts deserved closer scrutiny, but with the solar theory gaining the support of other Egyptology heavyweights, Breasted's views remained unchallenged for a long time.

He was fascinated by the mystery of the religion of the Ancient Egyptians. In his popular book, *The Development of Religion and Thought in Ancient Egypt*, he took it upon himself to show how, in his view, the development of Egyptian religious ideologies had occurred. The Pyramid Texts were the revamped product of 'successive editors almost at haphazard'.[20] 'What is the content of the Pyramid Texts?' he asked, and offered his wide and attentive audience this reply:

. . . it may be said to be, in the main, sixfold:
1) A funerary ritual and a ritual of mortuary offerings at the tomb
2) Magical charms
3) Very ancient ritual of worship
4) Ancient religious hymns
5) Fragments of old myths
6) Prayers and petitions on behalf of the dead king[21]

He reduced the Pyramid Texts to the mumbo-jumbo of archaic and superstitious magician-priests with weird ideas about the afterlife problems of their dead kings. Hardly a religion at all, put in those terms. True religious thoughts, Breasted believed, came much later, during the epoch of the 'heretic' pharaoh Ahkenaten (*c.* 1350BC).

By now, in Breasted's view, the solar cult was ready to become a solar faith with hints of a monotheistic concept. This was supposedly instigated by the new Aten cult introduced by the philosophical and gentle pharaoh, Akhenaten.[22] Breasted saw in Akhenaten's famous ancestor, the great Thoth-Moses III, a leader of a 'national priesthood as yet known in the early East, and the first Pontifex Maximus' under the god Amon. With Thoth-Moses III thus branded as a sort of pharaonic pope, whose office Breasted termed 'this Amonite papacy', his American audience began to conjure an almost Judeo-Christian idea of Thoth-Moses III's strange great-great-grandson. Much in Breasted terminology wishes to see Akhenaten as the precursor of a monotheistic religion with the sun, or rather sun disc, as the symbol of the One God, the 'Word'.[23]

This was not surprising to his audience, since Moses was believed by many to be a contemporary of Akhenaten and, some claimed, a main participant in the developing and blending of the monotheistic Hebraic faith with the religion of the pharaohs.[24] Nagging in the background, however, was the stellar cult which testified to Babylonian polytheistic star worship and was therefore unacceptable to Hebraic idealism. The stellar element was evident in the Pyramid Texts, and Breasted, as all others before him, felt uncomfortable with it.

He cast it as a half-baked theory which blemished the pure solar ideologies of the Pyramid Age.

Because of these flawed early studies, one of the most important keys to a true understanding of the texts – their use of allegorical astronomy – was nearly lost, buried under the mountain of academic verbiage which followed Maspero's publications. The astronomical key might have disappeared for ever but for a fateful discovery in 1982, a century after the Pyramid Texts were found. We will discuss this in later chapters, but let us now examine what the texts really are, and their relationship to the better known Egyptian Book of the Dead. This was a corpus of similar writings, recorded on papyrus scrolls in later times. Armed with this basic knowledge, we will be ready to approach the core of our mystery, the role of Orion in Egyptian religion.

III The 'Old Testament' of Ancient Egypt

We have seen that the Pyramid Texts are hieroglyphic writings carved on the internal walls of one of the Fifth Dynasty pyramids and four others from the Sixth Dynasty. They can thus be dated to a period between the earliest (Unas) *c.* 2300BC and the most recent (Pepi II) *c.* 2100BC. However, even these, the oldest religious writings in the world, are not the originals, but derive from some lost and more ancient archetype. We are fortunate in one respect though, that since the time when they were carved on the walls of the pyramids, they have not suffered from further corruption at the hands of editors and scribes, which cannot be said for other sacred scriptures from the distant past, including the Bible. It is sad that the Texts have been so neglected in recent decades by scholars of comparative religion and history of philosophy.

Considering the well-developed theology and mythology they contain, and the fact that they were used specifically for royal ceremonies and rites during the great epoch of the Pyramid Age, we can be sure that the copies which survived on the walls of the pyramids were indeed taken from older sources which have not themselves survived. How much earlier than the time of Unas was the original source material written?

Perhaps the best way to answer this question is to see how their discoverer, Gaston Maspero, and other scholars, Egyptologists and translators after him, perceived the Pyramid Texts. In a lecture Maspero gave soon after the discovery, he described them as '4000 lines of hymns and formulae, of which the greater part were originally written during the prehistoric period of Egypt'. Now 'prehistoric' Egypt, even by modern new chronological dating, places them around 3200BC at the latest – a date which Maspero and his contemporaries would have found very conservative indeed.

In 1912 Breasted was to write of these texts:

> Contrary to the popular and current impression, the most important body of sacred literature in Egypt is not the Book of the Dead, but much older literature which we now call the Pyramid Texts. These texts, preserved in the Fifth and Sixth Dynasty pyramids at Sakkara, form the oldest body of literature surviving from the ancient world and disclose to us the earliest chapter in the intellectual history of man as preserved to modern times.[25]

Since Breasted wrote those words, further confusion has been caused (particularly among investigators outside scholarly circles) by the common practice among Egyptologists of the first half of the twentieth century of referring to the funerary liturgy and many other texts of ancient Egypt collectively as 'The Book of the Dead', with the Pyramid Texts considered as the oldest version. This was a trend promulgated by, among others, Professor Wallis-Budge:

> The history of the great body of religious composition which form the *Book of the Dead* of the ancient Egyptians may conveniently be divided into four periods, which are represented by four versions:
> I. The version which was edited by the priests of the college of Annu (the On of the Bible, and

> the Heliopolis of the Greeks), and which was based upon a series of texts now lost . . . is known from five copies which are inscribed upon the walls and passages in the pyramids of kings of the Fifth and Sixth Dynasties at Sakkara, and sections of it are found inscribed upon tombs, sarcophagi, coffins, stelae and papyri from the Eleventh Dynasty to about AD200.[26]
> II. The Theban version, which was commonly written on papyri in hieroglyphics and was divided into sections or chapters, each of which had its distinct title but no definite place in the series. The version was much used from the Eighteenth to the Twentieth Dynasty.
> III. A version closely allied to the preceding version, which is found written on papyri in the hieractic character and also in hieroglyphics. In this version, which came into use about the Twentieth Dynasty, the chapters have no fixed order.
> IV. The so-called Saite version, in which, at some period anterior probably to the Twenty-sixth Dynasty, the chapters were arranged in a definite order. It is commonly written in hieroglyphics and in hieratic, and it was much used from the Twenty-sixth Dynasty to the end of the Ptolemaic period.

Budge's divisions are far from adequate. His versions II, III and IV, though similar to one another in many respects, differ markedly from the Pyramid Texts. Not only that, but the Pyramid Texts are lumped together with much later writings such as the Coffin Texts.

This banding together of Egyptian sacred writings and labelling them as 'Books of the Dead' has tended to cloud scholars' judgements concerning the Pyramid Texts and disguise their uniqueness. Budge did, however, go on to say that they 'bear within themselves proofs, not only of having been composed, but also of having been revised, or edited, long before the days of King Mena (*c.* 3300BC) . . .'[27] Dr Edwards, another former Keeper of Egyptian Antiquities at the British Museum and author of the definitive work on the pyramids of Egypt, reaffirmed this position when he wrote in 1947, 'For the most part the Pyramid Texts were not the invention of the Fifth or Sixth Dynasties, but had originated in earlier times . . .'[28] We can find no reason to doubt this assessment; indeed we believe that the Pyramid Texts and the star religion they contain predate the Fifth Dynasty by many centuries.

The final and definitive translation of the Pyramid Texts has proved an arduous business. After Maspero's hasty effort, German scholars were the most active in this field. Dr Kurt Sethe's epic version (1910–12) is foremost among them. During the 1950s and 1960s some English translations were to follow, the first by Samuel B. Mercer, Professor of Semitic Languages and Egyptology at Toronto University, then another by Alexander Piankoff based only on the Unas inscriptions.[29] Finally in 1969 the eminent and respected British philologist, Raymond Faulkner, produced what is considered the definitive translation. Published by Oxford University Press under the title *The Ancient Egyptian Pyramid Texts*,[30] Faulkner's translation is still regarded as the best. Eventually, in 1986, just over a century after their discovery, the publishers Aris & Phillips reissued Faulkner's book as the first paperback edition of the Texts, and this was reprinted in 1993. Faulkner, impressed by the antiquity and content of the inscriptions, described them thus:

> The Pyramid Texts . . . constitute the oldest corpus of Egyptian religious and funerary literature now extant. Furthermore they are the least corrupt of all such collections of funerary texts . . . They include very ancient texts among which were those nearly contemporary with the pyramids in which they were inscribed . . .'[31]

It is quite clear from all this that we are dealing with texts of which the greater portion originated well before the Fifth Dynasty. I felt safe, therefore, in assuming that, although the earliest copy was found in the pyramid of Unas, last king of the Fifth Dynasty, the Texts refer to a religion and rituals in existence during the Fourth Dynasty – the period during which the gigantic pyramids of Giza and Dashour were constructed.

In projecting the texts back one dynasty, from the Fifth to the Fourth, I believed I was not contradicting scholarly opinion. Indeed, all the Egyptologists involved with the Pyramid Texts, from their discoverer to their definitive translator, saw them as including very ancient material from beyond the Pyramid Age.

I was soon to discover, however, that while Egyptologists were prepared to agree to a greater antiquity for the Pyramid Texts than the Fifth Dynasty, they complained that there was no hard evidence of this. This seemed very odd to me; you could not have it both ways. Either it should be accepted, at least on philological grounds, that the texts contain very old ideas and material, or that they applied no earlier than the Fifth Dynasty. It was obvious that the ideas in the Pyramid Texts did not happen only during Unas's reign, and that they might have taken several centuries to develop into the royal state religion. Yet archaeologists wanted hard evidence, and that was not yet possible. Many scholars dismiss the philological evidence, which ought to be enough in such cases, and will not agree to the Texts being projected back before the time of Unas, not even to the Fourth Dynasty.

This paradoxical attitude created a scholarly impasse, which some more intrepid researchers have since challenged.[32] Many scholars preferred not to deal with the Pyramid Texts at all rather than risk embarking on the sort of controversy which could negatively affect their careers. The study of ancient texts was, it appeared, the *bête noire* of Egyptologists. Not many wanted to sink themselves in archaic texts said to be a 'haphazard' compilation of 'magical spells and hymns' of little or no consequence for the understanding of ancient ideas and 'sciences'. And anyway, enough had already been said about the Pyramid Texts by Breasted and others.

Regarding the projection backwards of the content of the Texts to earlier dynasties, or at least to the Fourth Dynasty, it is, in many ways, the same as saying that the Christian gospels (the earliest dating from the fourth century) should not be 'projected back' to the time of Jesus or even to the third century, when we know very well that Christianity was flourishing in both the east and in Rome. Unlike the prolific study of ancient Christian texts in hundreds of establishments around the world (not including the clerics), there is, as far as the Pyramid Texts are concerned, a curious academic seizure, a kind of intellectual catatonia which has struck many Egyptologists. A good example of this was expressed in a letter written to me by Professor Cathleen Keller, Senior Egyptologist at Berkeley University in California, who felt that a problem was raised by the fact that the versions of the Pyramid Texts which we possess date from the end of the Fifth Dynasty (at the earliest), somewhat later than the construction of the Giza monuments. She therefore thought that we should be cautious when projecting the texts back into the Fourth Dynasty.[33]

But Dr Keller did admit that 'many colleagues do not share this caution and frequently discuss the Giza complexes in terms of Pyramid Text rituals'. Yet what she did not make evident is just what is meant by 'some caution is called for'. I regarded a projection back to the Fourth as very cautious indeed, especially when it is recognised that the bulk of the Pyramid Texts in our possession were based on older originals.

Yet the well-known Professor of Egyptology, R. T. Rundle Clark, had warned in 1959 that 'Excessive caution leads to complete misunderstanding . . . It is in interpretation, however, that courage is needed.'[34] Here at last was an Egyptologist who was agreeing that 'the religious literature cannot be understood without some sympathy for the outlook of its authors'.[35] Rundle Clark saw the Pyramid Texts as the supreme achievements of their time and asked his colleagues to ensure that they were 'to be explained as such and not as a chance collection of heterogeneous tags put together to justify the pretensions of rival priesthoods'.[36] He emphasised that the more the texts are studied the greater appears their 'literary quality and intellectual content', and asked scholars to treat them with greater respect.

I soon realised what Rundle Clark meant when he said that courage is needed if you want to interpret the Pyramid Texts: the vague warning given by Dr Keller was nothing compared with a letter from a Swiss professor in Cairo who told me in no uncertain terms to leave things to the 'experts' and to go about my business. He advised me to 'abandon this subject and become a good engineer'.[37]

The more I investigated, the more it drew a mixed reaction from academics. Some felt that they could not comment on the 'mathematical' or 'astronomical' aspects of my thesis, others were nonplussed and most, at least in the early stage, simply could not be bothered to reply. I had the impression that not only was I treading on taboo territory, but that astronomy and the study of the Pyramid Age were anathema to Egyptologists: the two do not mix for them. Dr Keller summarised the problem when she wrote that many serious Egyptologists felt uncomfortable about the relationship between astronomical phenomena and ancient Egyptian architecture. They do not like to admit that the Ancient Egyptians were motivated less by scientific curiosity than by religious considerations in their understanding of the skies.[38]

The result of all this caution and antipathy about anything astronomical is that today, more than a century after the discovery of the Pyramid Texts, few non-specialist readers have even heard of them; fewer still are aware of the star religion or astronomies they contain.

We need now to re-examine what happened to the Texts after their discovery in 1880 and to explore their contents in the context of their allusions: the pyramid structures, the Nile Valley near Memphis and the sky above the two.

IV The Wrong Program for the Files

Anyone who has worked with a computer knows that calling up a file using a word-processing program not compatible with the one being used, means a garbled version of the text appearing on the screen.

This is more or less what happened (and in many ways is still happening) with the Pyramid Texts and the pyramids of

Egypt. We believe that the wrong program for reading them has been used. We are not talking of the translation from the hieroglyphic language to modern languages; we have the utmost faith in the work of Faulkner and others like him. We are referring specifically to the interpretation put on these texts by Egyptologists. We believe that the proper program or decoder exists and needs to be understood before we can properly decode the Pyramid Texts and extract their real, esoteric meaning. But let us first see how the orthodox consensus became established, and why it may be the result of using the wrong 'program'.

Although Maspero published large portions of the Pyramid Texts piecemeal from 1884 to 1894, these were distributed only among fellow scholars, as often happens with new archaeological finds of a textual nature. For example, the famous Dead Sea Scrolls, discovered in the 1940s, have only recently been published for the general public. Likewise, the Pyramid Texts were given little, if any, public exposure when they were first discovered. In 1910 Kurt Sethe produced the 'first standard edition'. This turned out to be a bulky work in three volumes which, apart from its high cost, was almost inaccessible for non-Egyptologists. (As a matter of interest, it was Sethe who coined the term 'utterance' to denote small chapters, sometimes only a few lines, in the main body of texts.)

The first sign of recognition that the star cult in the Pyramid Texts deserved closer attention came in 1946, when the prolific and tireless Dr Selim Hassan, an indigenous Egyptologist, gave his extensive interpretation of the Texts in a volume of his massive work entitled *Excavations at Giza*. Though Hassan was by no means in any mood, or position, to challenge Breasted's established solar contentions, he did pay far more attention to the stellar elements in the Pyramid Texts. He noted: 'At some remote period in the history of Egyptian religious thought, there was a belief that after death the soul of the King became a star among the stars of Heaven . . .'.[39]

Why Hassan saw this belief as being from 'some remote period' and not contemporary with the Pyramid Age is unclear. He drew his conclusions from what he read in the Pyramid

Texts and not, as his statement implies, from religious material from 'some remote period'. There is no religious material more remote than the Pyramid Texts. What Hassan meant was obvious: he saw in the Texts the elements of a star religion, but assumed that it came from a remote period because Breasted had said so. Breasted's reputation was now waxing in Egyptological annals, and his views had become academic dogma, not easy to dislodge. But the first crack in the solar theory was showing, and Hassan recognised that there were many references to the stars and the stellar destiny of pharaohs in the Pyramid Texts.

In 1952 Mercer produced the first English version of a manageable size and price. It came in four volumes, three of which were devoted to interpretations.[40] Mercer also paid more attention to the stellar doctrines of the Texts and, unlike Breasted and Hassan, began to recognise that hidden behind the liturgy was a primitive astronomy, expressed in poetic allegories and symbolism. He was perhaps the first to regard the Pyramid Texts as something other than a bulky compilation of 'hymns and spells' put together by some careless scribes. His analysis, though complex at times, was the first sign that someone was recognising in them elements of religious rituals which could be better understood through their stellar and astronomical content.

This, of course, conflicted with the established view, and Mercer was pilloried for being rash and far too bold in his interpretations. It was also said that his translations did 'not represent current knowledge of ancient Egyptian',[41] which was not entirely true. Mercer's study must have its place in the anthology of the Pyramid Texts, and his boldness may yet prove to be a good thing. (However, I soon discovered that quoting Mercer on the Texts was frowned on by academics.) He did much to highlight the fact that the Texts contain allegories about the stars and their movements and recognised that an astronomy mingled with mythology and rituals needed to be extracted from them. He showed that the principal theme was the powerful belief that the dead king would be reborn as a star and that his soul was believed to travel into the sky and become established in the starry world of Osiris-Orion, the god of the dead and of resurrection:

> The Dog Star was identified with Sirius; Orion was identified with Osiris. . . . It is not surprising to find an identification of Osiris with Orion . . . [for] one of the central themes of the Pyramid Texts was the complete identification of the dead king with Osiris . . .[42]

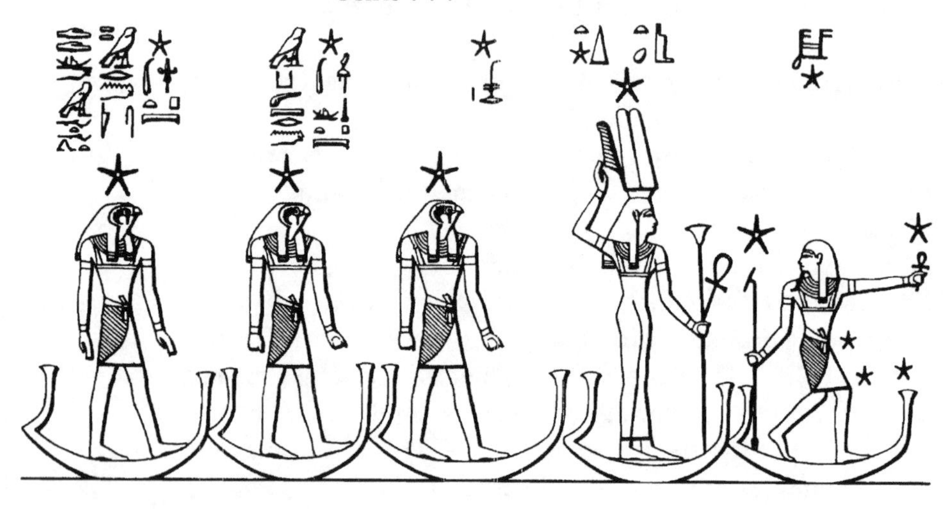

8. *Sahu-Orion followed by Sothis-Sirius and three stars in the Daily Procession of the Heavens*

Mercer also believed in the great antiquity of the cult found in the Texts: 'The worship of Osiris is, no doubt, prehistoric . . . by the time of the Pyramid Age it was a well-established cult'.[43]

The starry world of Osiris was called the *Duat*, and Faulkner, after the careful and meticulous analysis required to translate the Pyramid Texts, concluded that the Duat was not a part of the sun but often considered a 'part of the visible sky'.[44] Two years before he published his translation, Faulkner explored the star religion in the Texts and published his views in the prestigious *Journal of Near Eastern Studies*.[45] I am indebted to Dr Edwards for drawing my attention to this important article back in 1986 when the 'Orion Mystery' was still dragging its heels.[46] Faulkner quoted a large number of passages from the Pyramid Texts which mention the stars in connection with the soul of the dead kings and their afterlife destiny. Yet he ignored hundreds of other passages which also refer to the astral destiny of the kings, without specific reference to the word star, and more which drew attention to the stars by allegories and metaphors.

This is obvious from the way the dead king is identified with Osiris, who is identified with the constellation of Orion, as Mercer pointed out.[47] Faulkner also noted that the constellation of Orion was one of the afterlife dwelling-places of the souls of departed kings who became stars.

It was now becoming clear to me that observational astronomy and its material expression in the symbolic architecture of the pyramid structures and their orientations needed to be carefully examined. I discovered that I was not the only one who felt that a fresh review of the Pyramid Texts was imperative if progress were to be made in solving the mystery of the Egyptian pyramids.

The first serious complaint which called for new and unbiased review of the Pyramid Texts had come in 1948 from the eminent orientalist Dr Henri Frankfort, Professor of Oriental Archaeology at the University of Chicago and director of the Warburg Institute in London. Frankfort attacked Breasted's views as being 'biblical' and complained that no serious attempt was being made to extract the true meaning of the Pyramid Texts.[48] Two years after Mercer's publication of his commentaries on the Texts, a broadside came from another quarter, this time from a respected philological source. Alexander Piankoff, who had also translated part of the Pyramid Texts from the Unas pyramid, lamented:

> The approach to the study of Egyptian religion has passed without transition from one extreme to another. For the early Egyptologists this religion was highly mysterious and mystical. . . . Then came a sudden reaction: scholars lost all interest in the

> religion as such and viewed the religious texts merely as source material for their philological-historical research . . .[49]

In 1992, while Adrian and I were in the process of writing *The Orion Mystery*, another, more forceful call for a new appraisal of the Pyramid Texts – this time with due application of scientific astronomy – came from Jane B. Sellers, an Egyptologist who had been studying the astronomical contents of the Pyramid Texts for nearly sixty years.[50] In her recent book, *The Death of Gods in Ancient Egypt*,[51] Sellers airs the many complaints about how the Texts, and Egyptian religious texts in general, have been treated by scholars.[52] She quotes Henri Frankfort[53], who openly contested Breasted's stranglehold on the study of the Pyramid Texts:

> [James H. Breasted] described in 1912 a 'development of religion and thought in ancient Egypt' towards ethical ideals which pertained to biblical but not to ancient Egyptian religion. Since then interpretation (of the Pyramid Texts) has lagged . . . The most prolific writers . . . assumed towards our subject a scientist's rather than a scholar's attitude; while ostensibly concerned with religion, they were really absorbed in the task of bringing order to a confused mass of material.[54]

Sellers added a few comments of her own: 'Frankfort pointed out that men of this school have dominated the subject since the 1920s, and he accused them of both being responsible for the widely accepted view that religion was always a consequence of political power, and of being unable to see the wood for the trees.'[55]

Long before Jane Sellers's refreshing openness, I too had come to the conclusion that no one could really comprehend the Pyramid Texts by translating and interpreting the words without a background knowledge of observational astronomy. It was obvious that without this, and without a general appraisal of architectural symbology, they would remain unintelligible.

There could be no doubt that they were documents to be taken with the utmost seriousness and not be treated as the haphazard work of frivolous scribes. They showed evidence of being composed by a group of initiated priests-cum-astronomers who controlled the state religion of kings who were deemed gods and whose afterlife destiny was as established star souls in the world of Osiris.

But why build those massive pyramids to achieve this stellar destiny? What made them imagine that by taking the embalmed corpse of their king to 'his' pyramid in the Memphite Necropolis his soul would join Osiris in the sky?

4 LET THE PYRAMID TEXTS 'SPEAK'

There may be no need to try to connect the pyramid and the benben with the sun, as has often been done with unsatisfying effect, for the pyramid may be the agency for rebirth of the king, just as the decans (stars) themselves are reborn, as the Pyramid Texts say . . .
– E. C. Krupp, *In Search of Ancient Astronomies*

It is this mixture of astronomy and religion, the commingling of myth and reality, and this application of observing, engineering and surveying to the purposes of fantasy that so frustrates and fascinates the student of Egyptian life and science.
– James Cornell, *The First Stargazers*

I The Land of the Pharaohs

At the first opportunity, in autumn 1982, I took a short break and went on holiday to Egypt. Though I am of European extraction, Egypt is my native land and at that time my mother, also born in Egypt, was still living there. I am always revitalised by it: though poor in a material sense, it is rich in life and spirituality even today.

Alexandria, my home town, was once a great cosmopolitan city. Now dilapidated and overcrowded, it is bursting at the seams. It was named after its founder, Alexander the Great, and flourished under his successors, the Greek Ptolemies, to become a city that rivalled Athens and Rome for the beauty of its architecture and its location on the Mediterranean. Its fame as a centre of learning attracted philosophers and students from all over the Mediterranean world to its famous library and to listen to the liberal ideas and teachings of its platonic and pythagorian philosophers and advanced astronomers.[1] Under the Romans it remained a centre of learning and avant-garde ideas until the Arab conquest in the seventh century AD.

Alexandria had always been a city of ideas, a melting-pot of ethnic groups which included Greeks, Syrians, Ethiopians, Romans and Jews as well as the native Egyptians, known as Copts.[2] After the Arab conquest the city slowly fell into ruins as Egypt turned its back on Europe. It was to stay forgotten for many centuries until Napoleon invaded Egypt in 1798, but it was not until 1830, under Muhammad Ali, the first Turkish viceroy or khedive, that Alexandria began to regain some of its lost splendour. A keen, tough leader, Muhammad Ali invited Europeans – British, Maltese, French and Italians – to help him modernise Egypt, and within a century Alexandria was once more the most fashionable city of the Mediterranean. After the abdication of King Farouk in 1952, and the Suez War in 1956, pressure was put on foreigners by the Nasser regime, and Alexandria lost most of its Europeans, leaving it once again to the local Arabs. Unfortunately, the revolution could not solve the country's demographic problems, and Alexandria declined as Egypt's population grew alarmingly through the following decades. A country populated by only ten million in 1910 now has fifty-five million, increasing at a rate of one thousand a day. By 1982 Alexandria had become so crowded and dirty I could hardly recognise it as the city of my childhood.

As usual, a trip to the pyramids was on the agenda. I surmised that if any 'hard evidence' were to be found concerning the star religion of the ancients, it was here that one should look. After all, the pyramids were built at the time Robert Temple believed the star religion to have been of the greatest importance. Perhaps then the two were linked. Being trained as an engineer and surveyor, however, the evidence I was looking for would need to be more tangible than the interpretation of ancient myths. My lifelong experience of Africa and the Middle East made me especially sceptical of accounts by Dogon priests, however convincing they might appear. I wanted something physical, something you could see or touch and if possible measure. I was also wondering whether the Ancient Egyptians might have left some sign or message in the pyramids; otherwise why build them so large and of such robust construction? If a message *had* been left, surely it must have been concerned with their religious beliefs and might be the answer to the Sirius mystery I had read about. I was now looking for evidence of the 'first magnitude', the sort a specialist panel or jury would be compelled to accept.

On a warm night in May, two hours before dawn, I drove down from Alexandria to Cairo on the desert road. This poorly maintained road approaches Cairo from the north-west, so the first thing you see are the three pyramids of Giza. I arrived just in time to catch the light of the rising sun on their faces, their majestic presence inspiring awe and a sense of mystery. The site was free of the usual crowds of tourists; there were only a dozen or so keen visitors who, like me, were happy to miss a few hours of sleep to witness this magnificent sight.

I parked the car on a high spot overlooking the Giza plateau from the west, stood for a few moments to inhale the fresh morning air, then walked down towards the smallest of the three pyramids, that of Menkaura. A flutter of wings made me jump, and hundreds of pigeons and doves rose and circled the top of the pyramid. I had decided to climb a few stages to get a good photograph of the two larger pyramids against the light of the rising sun, and as I clambered up, I noticed that I was not alone. There, watching me nervously, was a small desert jackal. This was a rare sight as these animals, now nearly extinct in the environs of Cairo, are shy of humans. During all the years I had lived in Egypt, even on the many occasions when I had been out hunting in the western desert, I had never seen a jackal. This was a wonderful place and time for such a propitious encounter. We stared at each other for a few seconds, then the jackal disappeared around the corner. I suddenly remembered the discovery of the Pyramid Texts and how a jackal had led the *reis* to the entrance of the pyramid of Pepi I at Saqqara. No such luck here, I thought. There was nothing to suggest that soon I too would make a startling discovery about the Pyramid Texts: one that would alter the course of my life.

After sunrise I drove to Saqqara. I had not been there for many years and wanted to see again the famous inscriptions in the pyramid of Unas, last king of the Fifth Dynasty (*c.* 2350BC). The sun was now high in the sky and it was getting hot, so I stopped on the canal road and had breakfast. Arriving at Saqqara, I walked to the south side of the complex, avoiding the tourists and dragomen. Reaching the end of a long stone alley that had once been the symbolic causeway leading from the Nile to the pyramid complex, I could see the silhouette of Unas's pyramid. Viewed from the outside, it looks like a heap of rubble, but the same can be said of the other Fifth Dynasty pyramids. Yet Unas's pyramid is in many ways more precious than its perfect and gigantic predecessors. Unlike them, it is far from mute, for inside are the huge quantities of hieroglyphic texts.

An old *reis* in a shabby *jellabiyah*, the local garb, was guarding the entrance of the pyramid waiting for *bakshish* (a tip). A fiver in Egyptian currency, equal to two US dollars, makes you a VIP visitor; for fifty US dollars, the old man would wrap up the pyramid in a newspaper, if he could, and sell it to you. This is the sad state of Egyptian antiquities today. No one can blame the guardians of these monuments for trying to make the best of their situation; with dozens of mouths to feed and monthly wages that would not buy a meal in England, they rely on hand-outs from tourists by offering them 'privileged' access to the monuments. This often entails allowing tourists to touch the hieroglyphs, to use a flash camera and, if the *bakshish* is generous enough, to leave them alone in the monument to do as they please. Many of these men have been on the same job for decades, jealously guarding the richest territories along the main tourist routes, and some work without wages or pay a fee to have these lucrative posts. Over the years they have become my friends. They have learnt to love the monuments they are supposed to guard, albeit for different reasons, and given the right wages they would do a fine job.

Ibrahim, an old and tired *reis* I have known for years, was haggling with a noisy group of Japanese tourists. He gave me a broad smile and a *salaam* wave with his open palm and I did my usual recommendation act for him. I told the grinning Japanese how Ibrahim was once a 'friend of Howard Carter' and was said by 'Egyptologists' to be the best guide in the land. Then I urged them to give him a good *bakshish*, and asked them to make sure the ancient texts were not abused when they entered the pyramid. Leaving them nodding their heads in unison, I winked to the exalted Ibrahim and, slowly, bent my knees and lowered my head to enter the pyramid.

An awkward walk, more of a scramble, through a descending passage and then a horizontal corridor brought me into the first chamber where, like Maspero a century before, I looked at the limestone walls covered with carved texts. So well preserved are these that it is hard to believe that they were carved more than 4000 years ago. On the dimly lit wall the name 'Osiris-Unas' was written dozens of times in a neat row. Above it was 'Sahu', the ancient Egyptian name for Orion; then my eyes were drawn to the pitched ceiling covered with stars.

The Pyramid Texts, of which those from the Unas pyramid are the best examples, are uncorrupted by generations of editors and scribes. They are the original copy written on the stone more than 4000 years ago. It was these texts, the oldest known writings in the world, which confronted me now.

II Who Speaks for the Pyramid Texts?

Let the Pyramid Texts 'Speak'

One of the common problems concerning the study of ancient texts is that the appointed 'experts' will often not let the writings speak for themselves. They spend endless hours studying the contents and go through the material with a fine comb, but in the end many seem interested in using them only for philological studies and debates. In the course of this process, lacunae are filled in; simple words are replaced by complex ones; explanations, where they are given, are between brackets or sidelined into footnotes which draw the reader further into the morass of academic scaramouching. Nit-picking, and looking for flaws and technical errors in each other's arguments, causes more confusion than elucidation, and acts as a huge distraction.

The Pyramid Texts have not escaped this fate: a mass of scholarly verbiage has been thrown at them in the form of philosophical and philological arguments. Theological and etymological discussions have made their contents seem more esoteric than they need be. Decade upon decade of such treatment has reduced them to the status of boring material best left to the scholars and 'experts'. Thus the original texts, expressed in powerful terms which testify to a deep faith in an afterlife destiny, have been obscured.

Initially, I too fell into the trap of sieving through the articles and theses of academics, but it was apparent that some experts lacked any feeling for the texts, and spent their time contradicting and attacking one another. They presented the religion of the Ancient Egyptians as a bogus liturgy of rituals which made the rites of Roman Catholicism look straightforward.

There was only one way out of this impasse: I had to find the best translation available and make up my own mind about their meaning. I was able to get hold of Faulkner's acclaimed translation and begin with a clean slate. Our first rule is that wherever possible we should take passages at face value. Where possible the texts should be left to speak for themselves, and there are passages which speak plainly, even to a layman. It is only when this is done that we can hope to find the right connections between the texts and the material, visual aspect of the pyramid cult: the monumental architecture with its associated astronomy. When these two strands of evidence are considered together, we can understand the rituals of pharaonic rebirth.

However, the first question that must be tackled is whether the rebirth cult of the Ancient Egyptians was solar or stellar. In particular, did they believe that the departed king merged with the sun or was he supposed to become a star?

III The Star King of the Pyramid Age

Egyptologists have shown that the underlying concept of Ancient Egyptian theocracy was that while the king was alive he was a reincarnation of Horus, the first man-god king of Egypt, and was hailed as the son of Osiris and Isis. After his death it was believed that the pharaoh would depart to the sky and himself become 'an Osiris'.[3] But why an Osiris? What does the Osirianisation doctrine mean?

In Unas's pyramid the dozens of textual passages which call the dead king Osiris-Unas are emphatic declarations that, in his afterlife form, the mummified Unas was to be an Osiris. We are also told that the Osirianised kings became stars; not any stars but specific stars in the region of the constellation of Orion. Egyptologists thus concluded long ago that the rebirth ritual was essential to convert the dead kings into Osiris and more specifically (as Mercer argues for example) to Osiris in his astral form of Sahu, the constellation of Orion: 'Orion (Sah) was identified with Osiris . . .[4] It is not surprising to find an identification with Orion . . . [for] . . . one of the central themes in the Pyramid Texts was the complete identity of the dead king with Osiris . . .'.[5]

Central to the rebirth rites was that the dead Osiris was brought back to life through the magical rituals of mummification performed on him by his sister-wife, Isis, with the help

Let the Pyramid Texts 'Speak'

of Anubis. The importance of this idea was clearly understood by Jane Sellers, who says, 'the Pyramid Texts were aimed at ensuring the same rebirth for the dead king as that for the god Osiris-Orion'.[6] This is precisely what these texts are, a pharaonic 'life insurance' policy put there so that when the rebirth rituals were taken to the pyramid, the congregation could put into motion the magical words which would induce the soul of the dead king to become a star and rise to Osiris-Orion. Thus risen, the departed king would join the original Osiris and, like him, become a star god in the constellation of Orion. The original Osiris had become the Lord of the Duat, the realm of the dead inhabited by star beings.[7]

Dr Otto Neugebauer and Dr Richard Parker, who worked as a team for many years at Brown University in Rhode Island, and who were both acclaimed authorities on Ancient Egyptian astronomy, were the first to positively identify the sky image

9. *Ceiling from the Tomb of Senmut (New Kingdom)*
Sahu-Orion is shown with Orion's Belt above him.
He is preceded by the Hyades stars group and followed by Sirius-Sothis

of Sahu, seen as a huge human figure, with our own constellation of Orion.[8] They also noted: 'We know further from the names of the decans (star groups) of Sahu, "upper arm", "lower arm", etc., that Sahu was a human figure which in any case is graphically portrayed on the traverse strips of the coffin clocks and the various astronomical ceilings, such as Senmut's.'[9] In the Senmut ceiling a striding man can be seen with the three bright stars of Orion's Belt on top. Parker and Neugebauer correctly concluded that 'in the Pyramid Texts Sahu is identified with Osiris, which fits well with its depiction as a human figure on the coffins and ceilings'.[10] Many images of Osiris-Orion are shown in Ancient Egyptian drawings, among the oldest being that on the capstone or pyramidion of Amenemhet III's pyramid, in the Cairo Museum. Here, too, Sahu-Orion is seen as a striding man holding a large star in his hands.

It is clear from Egyptian funerary texts and the Pyramid Texts that Sahu-Orion was the soul of Osiris and that the sky region this bright constellation occupied was considered a very desirable place for the souls of kings to go to after the traumas of death and rebirth. Rundle Clark writes:

> The rising of Orion in the southern sky after the time of its invisibility is the sign . . . Osiris has been transformed into a 'living soul'. To achieve this, the second form of Osiris, for the deceased, is the basic purpose of the funeral rites . . . so as a new Osiris the dead king could, with due care by his successors, become one with the soul of the original Osiris.[11]

The first step in the astral transfiguration ritual was the changing of the corpse into an Osiris, i.e., the mummy-form. Thus to call the dead king, or rather his mummy, Osiris-Unas or Osiris-Pepi and so on, was to see the king ready to become a soul, that is a star in the Sahu-Orion region of the sky. This is made clear in the Pyramid Texts:

> O king, you are this Great Star, the Companion of Orion, who traverses the sky with Orion, who Navigates the (Duat) Netherworld with Osiris; you ascend from the east of the sky, being renewed in your due season, and rejuvenated in your due time. The sky has born you with Orion. . . . [PT882–3]

No interpretation is needed here. The texts state that the dead king becomes a star in Osiris-Orion. When this occurs is easily worked out, because we are told that the event is seen in the east, at dawn. This is confirmed by another passage:

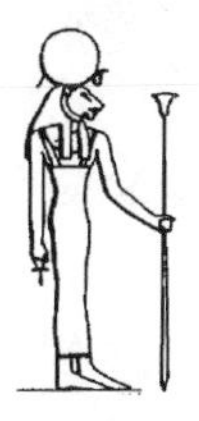

> Behold he has come as Orion, behold Osiris has come as Orion . . . O king, the sky conceives you with Orion, the dawn-light bears you with Orion . . . you will regularly ascend with Orion from the eastern region of the sky, you will regularly descend with Orion in the western region of the sky . . . your third is Sothis . . . [PT 820–2]

Faulkner, the definitive translator of the Pyramid Texts, used the Greek name of Sirius i.e. Sothis. From now on we shall refer to this star as Sirius in the astronomical context and as Sothis in the mythological context.

It is known that the star Sirius (Sothis) was linked to the start of the Nile's annual flood, which occurred around the end of June (mid-July in the Julian calendar). Sirius always rose immediately after the constellation of Orion and as such Isis, the goddess identified with Sirius, forms a pair or couple with Osiris-Orion. There are many such passages which mention Osiris-Orion and Isis-Sothis together, and many more which mention Osiris and Isis in their human form. Mercer seemed to think that when Sothis 'appeared as a goddess primarily, and not a star, she was represented as Isis . . . [and in this] . . . human form, she was closely associated with the constellation of Orion'.[12] This is easy to understand, for Sothis immediately follows Orion. Wallis-Budge said, 'the mention of Orion and Sothis is interesting, for it shows that at one time the primitive Egyptians believed that these stars were the homes of departed souls.'[13] The Pyramid Texts are categoric that the king becomes a star soul after death and, more specifically, joins Osiris-Orion in the sky. Many passages leave us with no doubt on this matter:

'The king is a Star . . .' [PT 1583]

'The King is a Star which illumines the Sky . . .' [PT 362, 1455]

' . . . The king, a Star brilliant and far-travelling . . . the king appears as a Star . . .' [PT 262]

'Lo, the king arises as this star which is on the underside of the sky . . .' [PT 347]

There can be little doubt that the Pyramid Texts make a clear statement that the dead kings become stars, especially seen in the lower eastern sky. They also tell us that it is the souls of departed kings which become stars:

'be a soul as a living star . . .' [PT 904]

'I am a soul . . . I (am) a star of gold . . .' [PT 886–9]

'O king, you are this great star, the companion of Orion . . .' [PT 882]

' . . . behold he (the king) has come as Orion, Behold Osiris has come as Orion . . .' [PT 820]

Thus the dead king was an Osiris and his soul was an Osiris soul, whose depiction in the sky was Orion. The Pyramid Texts call the starry afterworld of Osiris the Duat, and it is in this Duat region that the astral souls become established. There are many indications that the Duat included the constellation of Osiris-Orion and that it was also thought of as the pyramid fields in the Memphite Necropolis:

'The king has come that he may glorify Orion, that he may set Osiris at the Head . . .' [PT 925]

'The Duat has grasped your hand at the place where Orion is . . .' [PT 802]

'May you ascend to the sky, may the sky give birth to you like Orion . . .' [PT 2116]

'Live and be young beside your father (Osiris), beside Orion in the sky . . .' [PT 2180]

'In your name of Dweller in Orion . . .' [PT 186]

'O king, you are this Great Star, the companion of Orion, who traverses the sky with Orion, who navigates the Duat with Osiris . . .' [PT 882]

The departed Osiris-king was to join Osiris-Orion in the prescribed region of the sky, where all other departed kings, (the royal ancestors) had gone. We can even gauge the time of year considered ideal for this astral rebirth ritual: We are to consider Orion's rising at dawn, but we are also told that Sothis is involved, so this star must also be visible at dawn. We also know that this was the prelude to the start of the annual flood of the Nile which occurred near the summer solstice. For the three events to occur at the same time during the Pyramid Age, astronomical calculations give the date of *c.* 2750BC.[14] A passage in the Pyramid Texts alludes to this ideal time:

The reed-floats of the sky are set in place for me, that I may cross by means of them to Ra (the rising sun) at the horizon. I ferry across that I may stand on the east side of the sky, when [Ra] is in [his] northern region among the imperishable stars, who stand at their staffs and sit at their east . . . I will stand among them, for the Moon is my brother, the Morning Star is my offspring . . .' [PT 1000–1].

The words in square brackets have been inserted to give the

correct astronomical sense to the passage. Near the summer solstice the sun is 'in the northern region' of the sky, and rises at azimuth 63.5 degrees, that is some 26.5 degrees north of due east.[15] At this time Orion rises just a few degrees south of due east so that the king 'may stand on the east side of the sky'. Using a special computer program to recreate the dawn sky for *c.* 2750BC at dawn on the summer solstice, we get a visual picture of the ancient textual description. Orion is 'fully risen', and this all-important moment is denoted by the appearance of the 'Star of Isis', Sothis, just over the horizon. It was exactly then that the bright star, Isis-Sothis, performed its heliacal or first dawn-rising to mark a 'new birth' and the beginning of a new year.[16]

IV Offspring of Isis-Sothis and Osiris-Orion

Although debate is rife among Egyptologists as to how the Pyramid Texts and the rituals they present should be considered, they are unanimous on one thing: the royal rebirth rituals were based on a dramatic re-enactment of the Osiris and Isis story and the miraculous seeding and subsequent birth of their son and heir, Horus.

Nowhere is the Osirian myth given in full form; it seems that the Ancient Egyptians knew it so well it was deemed unnecessary to narrate it as a preamble to the rituals, just as the majority of Christians know the basic elements of Christ's story. There are, however, thousands upon thousands of references to Osiris, Isis and Horus in the Ancient Egyptian funerary texts, including the Pyramid Texts, so it has not been difficult for Egyptologists to reconstruct the Osiris story:

> Osiris was the eldest son of Nut, the sky goddess, her other children being Isis, Seth, Nephthys and possibly Anubis. Osiris, a man as well as a god, became the first king of Egypt and his sister Isis became his consort. He was a good king and estab-

lished the rule of law (*maat*). With the help of his vizier, the 'god' Thoth, he taught men religion and the arts of civilisation. Egypt became prosperous and it was at peace with itself. Unfortunately, not everyone was happy – especially his brother Seth. He plotted against Osiris, murdered him and cut up the body into small pieces, which he

10. Scenes from the Book of the Dead, the weighing of the heart and presentation of a worthy soul at the court of Osiris. Osiris is attended by his sister-wife Isis and their sister Nephthys. Before him are the four sons of Horus, standing on a lotus

scattered all over Egypt. Even more tragically, Isis was still childless when this happened and Osiris had no heir to take his place. All was not lost, however, for Isis secretly gathered up the pieces of her husband's body and, by means of her magical powers, reconstituted them into the body of Osiris, thus making the first-ever mummy. Having brought him back to life, she was now able to have sex with him. Although this was only a temporary reprieve for Osiris, it was long enough for Isis to become pregnant with his seed. His task on earth having been completed, Osiris transfigured himself into a star being (Orion) and went on to rule the Heavenly Kingdom of the Dead – called the Duat. Isis now hid from Seth in the marshes of the Delta near Heliopolis and in due course gave birth to a son, Horus. He grew up to become a powerful prince, and eventually challenged Seth to a duel to see who had the right to rule Egypt in Osiris's stead. During the fight, Horus lost an eye and Seth lost his testicles. Though the battle was inconclusive, the sun god was eventually persuaded to judge in favour of the young Horus and he was proclaimed king, the first in the line of the pharaohs.[17]

The tragic story of Osiris and the heroic struggle of Horus to regain the throne served as a model throughout Egyptian history. The pharaohs legitimised their authority and, more especially, deified their rule by proclaiming themselves reincarnations of Horus; the epic battle with Seth became a metaphor for the struggle of the pharaoh against illegitimate claims to the throne. It is an accepted fact that all kings of Egypt were regarded as the reincarnation of Horus, and in this capacity they were the upholders of what the ancient Egyptians called *maat*,[18] or 'law and order'. When a Horus-king died, he was assured a rebirth with Osiris, that is to say he became at one with Osiris in the afterworld of the Duat. This would leave the throne of Egypt vacant for the legitimate heir to assume the role of Horus; the heir was thus the living one, the son of Osiris and Isis, as opposed to his dead father, now an Osiris-king. It was this cyclical exchange from 'Horus-to-Osiris-to-Horus' which was at the heart of the royal cult of the pharaohs; being gods, their mortality could be explained only in terms of this divine myth and it was never in doubt that they would be reborn in the afterworld realm of Osiris. The essential aim of the Pyramid Texts was to assist in this crucial process.

Henri Frankfort showed that the rebirth rites for a dead king ran in parallel with the coronation rituals for his heir.[19] The death of a pharaoh thus triggered a double event, his funeral and the coronation of his heir, and we should be aware, when reading the Pyramid Texts, that we are dealing with a double ritual: the funeral of a Horus-king waiting to become an Osiris, and the coronation of the new Horus-king as the son of Osiris. In astral terms, the new king was the son of Osiris-Orion. Just as Osiris was identified with the constellation of Orion, so his consort and sister, Isis, was identified with Sothis (Sirius). Isis-Sirius (Isis-Sothis) was thus the astral mother of the living king. Sirius, as we have seen, is the brightest star in the sky and its constellation, Canis Major, immediately follows Orion in its rising.

In the Pyramid Texts the living king, the new Horus-king undergoing his coronation while attending to his father's rebirth, makes these evocative claims:

'"How lovely to see", says she, namely Isis . . . to my father, to the [dead] king, when he ascends to the sky among the stars . . .' [PT 939]

'The sky is clear, Sothis [Sirius] lives [appears], I am a [the] living one, the Son of Sothis . . .' [PT 458]

'Your sister Isis comes to you rejoicing for love of you. You [the dead king] have placed her on your phallus and your seed issued in her, she being ready as Sothis, and Har-Sopd has come forth from you as Horus who is in Sothis . . . and he (I) protect(s) you in his (my) name of Horus, the son who protects his father . . .' [PT 632–3]

'The (dead) king's sister is Sothis, the king's offspring is the Morning Star . . .' [PT 357; 929; 935; 1707]

The dead Osiris-king also makes his claims:

'The sky is pregnant of wine (the dawn light), Nut has given birth to her daughter (Sirius) [in] the dawn-light, I raise myself indeed . . . my third is Sothis . . . [PT 1082–3] [the second here being the offspring'.]

'Give command to him who has life (i.e., the living king as Horus), the Son of Sothis, that he may speak on my behalf and establish my seat in the sky' [PT 1482]

These extracts indicate clearly the performance of an evocative stellar ritual, in which the dead king, as a star of Osiris-Orion, is seen as copulating with Isis-Sothis (Sirius) to seed her womb and leave her pregnant with the astral Horus, the son of Sothis. The latter is represented by the legitimate heir, now to become the new pharaoh of Egypt. It seems obvious that this son of Sothis is also identified with a celestial body, and Faulkner has suggested the planet Venus (a star) because of the name 'Morning Star'.[20] But neither Venus nor any other planet qualifies as 'Morning Star' so that it is also 'coming forth' from the womb of Sothis (Sirius). Who or what was the 'Morning Star' supposedly close to Sirius?

In the epoch of *c.* 2750BC, Sirius had a declination of about −21.5 degrees.[21] This caused it to rise quite far off the ecliptic during the summer solstice, at about azimuth 116.5 degrees or some 26.5 degrees south of east, with the sun being about 54 degrees away to the north just below the horizon. This means that none of the planets could be anywhere near Sirius during its heliacal rising. So what bright star could be called the 'Morning Star' and considered so close to Isis-Sirius? Was there a bright object near Sirius which the ancients saw but which has now become invisible? Is a 'lost' star a real possibility?

Here we must recall Robert Temple's *Sirius Mystery*, the heart of which revolves around the secret knowledge of the Dogon of Mali, who reported an invisible companion star of Sirius. According to Temple, the Dogon's tradition supposedly came from Ancient Egypt, where it originated *c.* 3200BC. Today this invisible star is called Sirius B. It is super-dense, a white dwarf in astronomers' jargon, which canbe seen only through a very powerful telescope. Scientists do not think that Sirius B was visible in ancient times, but could they be wrong?

But let's leave this controversy for now, while we develop the Orion Mystery further and look at the mysterious shafts in the Great Pyramid.

V Channels to the Stars

In the Great Pyramid are four protracted and narrow channels or shafts which have long baffled Egyptologists. We discussed them briefly earlier in this book, but now we need to return to them in greater detail.

The two shafts within the King's Chamber had been known since the early seventeenth century. John Greaves, Savillian Professor of Astronomy at Oxford, reported the existence of the openings of these channels when he made his famous survey of the Giza pyramids in 1638, and noted that the northern one was blackened by 'lamps burning there'.[22] De Maillet, the French consul-general, also reported the shafts in 1693 but came to the odd conclusion that they had been used to lower food and clear detritus during the construction of the pyramid.[23] Jomard, who accompanied Napoleon to Egypt in 1798, later described 'these deep narrow cavities which emanate from the walls of the central chamber' in Khufu's pyramid.[24] It was the British adventurer, Colonel Vyse, and his colleague, J. S. Perring, who discovered the outside openings of the shafts of the King's Chamber in 1837. At first they thought the shafts led to a room, despite the small cross-section (about 22 × 23 centimetres), but abandoned this idea when the air rushed through the chamber after they cleared the southern shaft.[25] They then decided, erroneously, that the shafts had been designed for ventilation, and coined the term air-shafts. Flinders Petrie accepted this conclusion and adopted the term air channels in his description, saying 'the air channels leading from this [the King's] chamber were measured on the outside of the pyramid; the north one varies from 30 degrees 43 minutes to 32 degrees 4 minutes in the outer 30 feet; the south one varies from 44 degrees 26 minutes to 45 degrees 30 minutes in the outer 70 feet.'[26] In 1872, Waynman Dixon, a British engineer, conjectured that similar shafts might be found in the Queen's Chamber, lower down the monument. Piazzi Smyth, the Astronomer Royal of Scotland who employed Dixon, explains how the discovery was made:

> Perceiving a crack (first I am told, pointed out by Dr Grant) in the south wall of the Queen's Chamber, which allowed him at one place to push in a wire to a most unconscionable length, Mr Waynman Dixon set his carpenter man-of-all-work, by name Bill Grundy, to jump a hole with hammer and steel chisel at that place. . . . next measuring off a similar position on the north wall, Mr Dixon set the invaluable Bill Grundy to work there again with his hammer and steel chisel . . .[27] [Smyth, P. *The Great Pyramid*, p. 428]

Sir Flinders Petrie who measured the slopes of the shafts in 1880 explains how this was done:

> The channels leading from this [the Queen's] chamber were measured by goniometer; they are exactly like the air channels in the King's Chamber in their appearance, but were covered over the mouth by a plate of stone, left not cut through in the chamber wall; no outer end has yet been found for either of them, though searched for by Mr Waynman Dixon, who first discovered them, and also by myself . . .

But then followed an odd commentary by Petrie: 'I observed something like a mouth of a hole in the 85th course on the south face, scanning it with a telescope from below; but I was hindered from examining it closely . . .'[28]

We know, from Gantenbrink's recent extensive survey using alpine gear, that when his team scanned the north and south faces for the alleged openings, Petrie was wrong about seeing the mouth of a hole on the south face. Neither of the two shafts in the Queen's Chamber pierces the pyramid to the outside. Egyptologists later claimed, wrongly, that these shafts stopped some eight metres from the walls of the Queen's Chamber. Petrie gave the mean slopes of these shafts as, north channel 37 degrees 28 minutes and south channel 38 degrees 28 minutes, each statement being 'the mean of two observations, which never differed more than six minutes [Arcminutes]'. Petrie was also to be proved wrong. However, the implications of his report were tremendous, for it did much to divert attention away from the air channels in the Queen's Chamber; the logic being that since they did not pierce the pyramid, the channels (and consequently the Queen's Chamber) were abandoned by the ancient builders in favour of the King's Chamber, higher up the pyramid. This idea persisted for many decades until Rudolf made his discoveries in the southern channel of the Queen's Chamber, showing that it was cut much deeper than Petrie had deduced, and extended well above the floor level of the King's Chamber, about 19.5 metres higher, thus running almost parallel to the southern shaft of the King's Chamber for the last 25 metres of its track.[29]

The ventilation theory had long been questioned; in 1924 a Belgian Egyptologist, Capart, suggested another plausible function for the shafts. Sensitive to the symbolic function of the whole monument, Capart did not think they were air-shafts at all, but served a religious purpose instead: 'it is more probable that they had a funerary purpose, perhaps to afford a passage for the soul of the king.'[30] The same idea was expressed by the German Egyptologist, Steindorff, in 1929,[31] cautiously by Edwards in 1947[32] and by Vandier in 1954.[33] About the same time as Vandier, the symbolic function attributed by Capart to the shafts was investigated more closely by Badawy, an Egyptologist with a knowledge of Egyptian architecture. A breakthrough was on the way.

In his detailed work on Ancient Egyptian architecture, Badawy suggested that the shafts in the King's Chamber could have served as channels to the stars, 'the northern passage . . . for the voyage of the soul to the imperishable circumpolar stars, the southern one to Orion.'[34] So entrenched was the idea that the Pyramid Texts reflected a solar destiny for the dead king that no one had thought of this. It was not until 1964 that Badawy sought the help of an astronomer to make the precessional calculations which would validate his theory (see Appendix 1). He asked Virginia Trimble to help him with the problem, and they jointly published their work in an Egyptological journal in Germany.[35] Badawy first considered the view that the shafts had been intended for ventilation:

> This interpretation does not . . . withstand objective criticism. Besides the fact that no provision was ever made by the Egyptians in any of their various types of tombs this one, if so interpreted, would conform but poorly with their achievement in ventilating their houses.[36]

Badawy's architectural studies had shown that the Ancient Egyptians did not ventilate tombs, nor would one expect it.[37] As for the ventilation of their houses, they used slanting channels opening on the ceiling and oriented north to make use of the cool northern breeze. Badawy correctly pointed out:

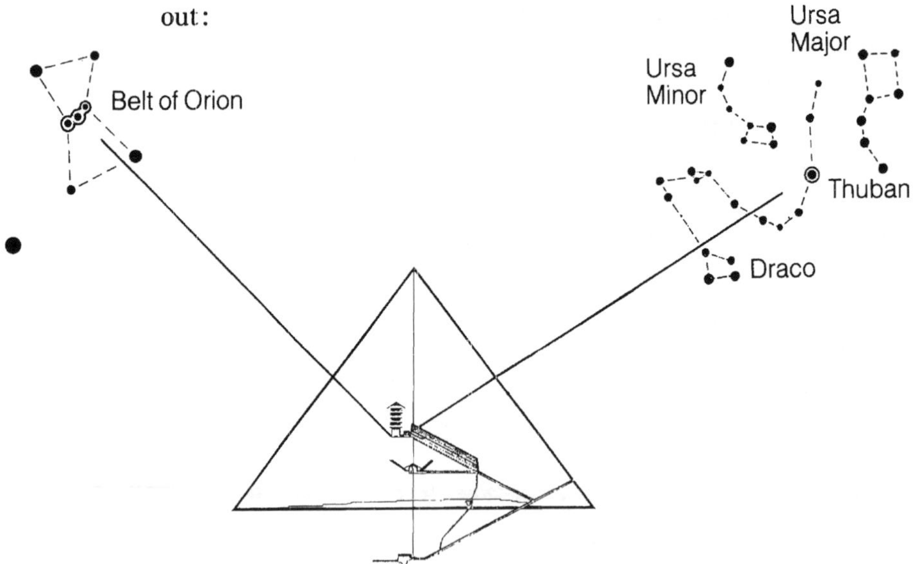

11. *The Great Pyramid of Giza in cross-section*
Alignments of shafts to stars c.2600BC as discovered by A. Badawy and V. Trimble in 1964

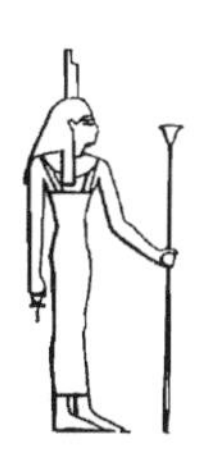

> To ventilate the burial chamber of Cheops channels running horizontally at the level of the ceiling would have been more adequate than the inclined shafts that start at about one metre from the floor, at the level of the lid of the sarcophagus. One should add to this inadequacy in the design all the constructional problems involved of the building of the two inclined shafts through all the courses, a process which could have been avoided by building them through one horizontal course.[38]

He also pointed out that the opening of the shafts in the Queen's Chamber had been left uncut in the walls, and that it was quite likely that the same had applied to those in the King's Chamber; if this was the case, 'their assumed purpose for ventilation would have been out of the question'.[39] Badawy knew, of course, that the Pyramid Texts referred to Sahu-Orion and that the departed Osiris-king was identified with these stars. Orion has always been a southern constellation and so it seemed an obvious goal to consider for the southern shaft of the King's Chamber. The average slopes were taken by Badawy from Petrie's data, around 44.5 degrees for the southern shaft and 31 degrees for the northern. It was immediately obvious to a trained astronomer like Trimble that the northern shaft pointed close to the celestial pole, which lies at an altitude of nearly 30 degrees as observed from Giza – 29 degrees 58 minutes 51 seconds to be exact as measured from the centre of the Great Pyramid. Virginia Trimble worked out the declination of the stars of Orion's Belt for *c*. 2600BC, then the assumed date for the Great Pyramid. She obtained the results shown in the first table.[40]

Orion's Belt	Declination in 2600BC	
Al Nitak (Zeta Orionis)	−15°	33′
Al Nilam (Epsilon Orionis)	−15°	16′
Mintaka (Delta Orionis)	−14°	45′

Source: Appendix 1.

The exact latitude to the nearest minute of arc given for Cheops's pyramid is 29 degrees 59 minutes. At this latitude the celestial equator, an imaginary line dividing the northern and southern hemispheres of the apparent sky globe that encompasses the earth, lies at an altitude of 60 degrees 01 minutes above the southern horizon i.e., the meridian looking south (90 degrees − 29 degrees 59 minutes = 60 degrees 01 minutes).

The celestial equator is taken as being zero declination, so that anything above it is a positive declination in the northern hemisphere of the sky, and anything below it is a negative declination in the southern hemisphere. To work out the altitude of a star at the meridian as seen from Giza and looking south, the declination has to be subtracted from the altitude of the celestial equator (60 degrees 01 minutes):

Orion's Belt	Altitude, degrees and minutes
Al Nitak (Zeta Orionis)	$(60° 02' - 15° 33') = 44° 29'$
Al Nilam (Epsilon Orionis)	$(60° 02' - 15° 16') = 44° 46'$
Mintaka (Delta Orionis)	$(60° 02' - 14° 45') = 45° 17'$

Both Trimble and Badawy quickly realised that it was no coincidence that the southern shaft, which pointed towards the meridian at a slope of 44 degrees 30 minutes, seemed to target the passage of Orion's Belt. Trimble also showed that no other important stars at that epoch passed at this point in the sky: 'It would seem likely that some other stars might pass in the same fashion over the opening of the shaft. It happens, however, that no other stars of comparable magnitude had declinations within 1 degree 30 minutes of −14 degrees 30 minutes during that period'.[41]

Badawy thus concluded that this shaft was aimed *deliberately* at Orion's Belt, the centre of the Sahu-Osiris constellation, to help the soul of the dead king to rise to the special starry heaven of Sahu-Osiris (Orion). Badawy was actually 0.5 degrees out, for we now know that the southern shaft of the King's Chamber is at 45 degrees. But 44 degrees 30 minutes was close enough for Badawy and Trimble to make this startling revelation. Oddly, neither Badawy nor Trimble pursued the same logic with the two shafts in the Queen's Chamber, perhaps accepting the consensus that these had been abandoned.

The discovery that the southern shaft of the King's Chamber was targeted in *c.* 2600BC to the three stars of Orion's Belt was largely ignored at the time. Only Edwards took the matter up, but not until 1981, when he made these important comments in an article written in honour of his American friend, Dows Dunham:

> The Pyramid Texts frequently allude to the king's association in his afterlife with the stars and, in particular with the circumpolar stars and with Orion and Sothis. Scientific study has shown that the northern channel (shaft), which sloped upward at an angle of 31 with the horizontal, was almost in exact alignment with what was then the Pole Star (alpha Draconis), while three stars in Orion's Belt passed each day at culmination directly over the southern channel (shaft), whose slope is 44.5. To suppose that such a setting of the channels had no magical significance seems highly improbable.[42]

It is strange that no Egyptologists have taken Badawy's work further; perhaps because it challenges the theory of a solar destiny for the king, which still dominates pyramid studies. Yet it should be obvious that the orientation of this shaft towards the Belt of Orion was connected with the many statements in the Pyramid Texts that the afterlife destiny of the pharaoh was in that region of the sky.

When I began my investigations into the star religion of the pharaohs, I knew nothing of Badawy or his article; had I known, it might have saved a great deal of time and effort. More importantly, it might have given me the encouragement not forthcoming from Egyptologists at the start of my quest. As it was, I turned my attention to Giza, unaware of these vital stellar clues.

THE GIZA PLAN 5

They [the builders] were apparently able to dictate . . . the small dimensions of the Third Pyramid, despite the presumed desire of Menkaura [Myцerinos] to have a monument equal to those of his predecessors . . .

– J. A. R. Legon in *Discussions In Egyptology*

At Giza we are confronted by a set of monuments which bear every sign of intelligent design, yet we are ignorant of the principles upon which these designs were based.

– R. Cook, *The Pyramids of Giza*

I A Peculiar Offset

In 1982, the day after I had visited the pyramid of Unas, I went to another familiar haunt, the Cairo Museum of Egyptian Antiquities. My objective was the east wing of the ground floor, where most of the Pyramid Age relics were kept.

The Museum is an extraordinary place, in the heart of Cairo on the bustling north side of Tahrir Square, and entering its courtyard is like finding a sanctuary from the madness of the traffic outside. The present building was designed at the turn of the century by the French architect, Marcel Dourgnon; not unexpectedly there is a distinctly French feel about the place, due not least to the mausoleum of Mariette. He had asked that his remains be entombed in a sarcophagus and kept in the gardens of the Museum; recently his statue has been repaired and now dominates the crowds as they stream into what was once his exclusive domain. In a country becoming more fundamental by the hour, Mariette's statue looks strangely out of place, a relic of a colonial past Egyptians would have preferred to do without. The gardens at the entrance of the Museum are full of pharaonic relics, which would receive pride of place anywhere else. There is no more room inside the building so many statues and sarcophagi are left to the mercy of the city's terrible pollution and the groping fingers of thousands of tourists. On the east side of the Museum is a local school, where two sarcophagi serve as school benches and a third as a rubbish bin.

I walked through the main hall and made my way to the pyramidion (benben) of Amenemhet III's pyramid. This is dated to *c.* 1850BC and once stood on top of the king's now crumbled pyramid at Dashour.[1] It is made of highly polished black granite and has two lines of inscriptions around its base. These, as well as the winged disc and eyes of the Horus symbol, include the figure of Osiris-Sahu (Orion) with a star in his outstretched hand. Before entering the east gallery which contains the Old Kingdom relics, I came across a statue of Menkaura, builder of the third pyramid of Giza. Though small in size, it is beautifully cut from green schist. The king seems to radiate a powerful authority and a strange intensity of feeling, very characteristic of Old Kingdom statuary art. Menkaura is presented on each side by a goddess; both display an odd sense of tenderness mingled with pride in the way they hold on to Menkaura's arms. Clearly, such kings were not the tyrants they are sometimes made out but were regarded as deified rulers to be loved and glorified.

Passing into the famous Room 42 which contains many Fourth Dynasty relics, I immediately saw the splendid statue of Khafra; builder of the second pyramid. This is cut from a single block of black diorite, a granite stone which is extremely hard to work. Yet the statue is so finely polished that it looks like metal; it is considered by some as one of the world's great works of art. The sculptor who worked the stone must have been the Michelangelo of his time; how he sculpted diorite to such perfection with only copper tools remains a puzzle. Khafra sits on a throne, his face radiating both authority and love, depending from which side you observe it. His head is embraced by the open wings of a Horus hawk which rests on Khafra's shoulders. I felt that even the ornate and beautiful relics from Tutankhamun's tomb did not have such haunting beauty.

I walked around the gallery taking in as many impressions as I could, then my eyes caught something else: a large poster on the north wall – an aerial photograph of the Giza pyramids. The tag indicated that it had been taken by the Egyptian air force, probably in the 1950s, and was the first aerial view I had seen from directly above the Giza site. Before looking at this poster, I had not paid much attention to the curious offset of Menkaura's pyramid from the south-west alignment of the two larger pyramids. But now, looking from high up over the site, it stuck out like an out-of-plumb frame on a wall. I had worked as a setting-out building engineer a few years before,[2] and my eyes were trained to focus on such anomalies in site layouts. The pyramid of Menkaura, I felt, *was not quite where it ought to be*. I asked the guard if I could take a photograph of the poster and got a smile and nod, with the usual military style salute which indicated *bakshish* was in order. I was using a black-and-white film with a fast 50mm lens on my old Olympus. I raised the camera and clicked only once. This idle snapshot was to change the course of my life.

The short holiday over, I returned to work in Saudi Arabia. In Riyadh I had the film developed and ordered several large copies of the aerial photograph of the Giza pyramids. I was intrigued by the offset of Menkaura's pyramid and wanted to try to solve the riddle. Most of my friends in Saudi were in the construction industry – civil engineers, architects, planners –

and I felt that their opinion might help. My aim was to see if we could agree on the reasons for the odd layout plan of the three pyramids.

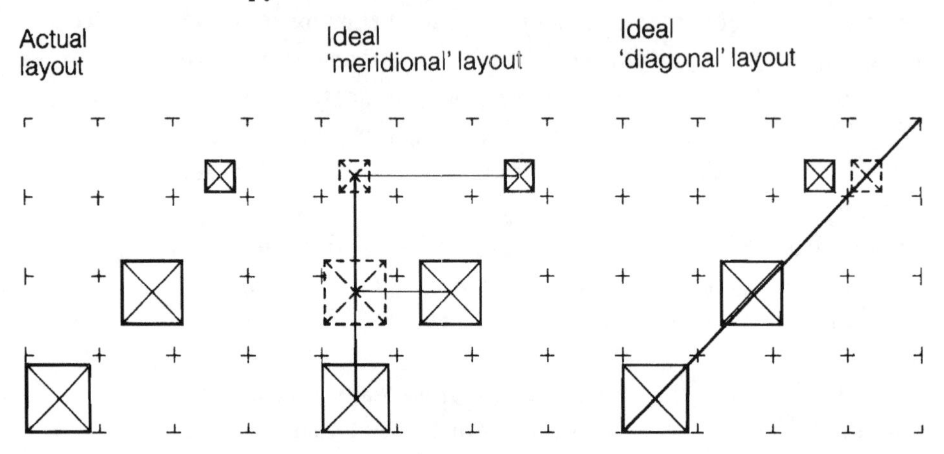

12. *Analysis of the Giza Layout Plan*

As I had thought, most of those who looked at the photograph made the same observation: the three pyramids were each set along their own meridian (north-south) axes and everyone noticed the south-west diagonal along which the two larger pyramids are set. They agreed that this indicated a unified plan. Then came the confusion I had anticipated: they wondered why the third pyramid was so much smaller than the other two, and, even more puzzling, why it was slightly offset east of the south-west diagonal line which linked the two larger pyramids. All agreed that the size and offset of the Menkaura pyramid had been a deliberate choice by the architect. The question was why?

II An Architectural Plan

I decided to give copies of the picture to another group of friends not involved in construction work, but with some understanding of an artistic or a poetic nature. I wanted to see if they would ask the same questions. This time, however, I traced in black ink the south-west diagonal line which linked the two larger pyramids, and extended the line to the Menkaura pyramid to show the curious offset. I also provided them with the assumed sequence of building: Khufu (Cheops with the Great Pyramid), then Khafra (Chephren with the second pyramid) and Menkaura (Mycerinos with the smaller third pyramid). I was listening for questions as to why this last pyramid was much smaller and was offset from the south-west axis of the other two larger ones. The replies confirmed what the other group had deduced: the size and offset of Menkaura's pyramid seemed a deliberate choice by the architect. This group, however, was more concerned to discover why Menkaura's pyramid was so much smaller. I dished out the standard reply: Menkaura probably was short of resources. They were not satisfied, nor was I, but it was as good an explanation as any. Yet on what evidence did this lack-of-resources hypothesis rest? As far as I could make out, there was none. Egyptologists think that Menkaura ruled for just as long as his two predecessors at Giza and was described as an equal to Khufu and Khafra, his eternal cronies. I offered another standard answer: Menkaura was in a hurry, so he built a smaller pyramid. Again I had to agree that there was really no evidence for this conclusion. The pyramid must have taken several years to build – seven to ten years on conservative estimates[3] – so how could Menkaura have been in a hurry? Was he a sick man? Again no evidence. His statues show him as healthy and strong.

I saw the point that this second group was trying to make: whichever way you looked at it, it didn't make sense for Menkaura to have settled for a much smaller pyramid. He most likely had the same autocratic power and resources as his immediate predecessors. In any case, the concept of economy was alien to them; resources meant plenty of able men and plenty of limestone quarries, both of which Menkaura certainly had. His predecessors would also have left much behind to make his task easier: ready open quarries, tools, accommodation for workers, sledges and so forth (what construction companies today call the 'preliminaries'), and a wealth of experience gained by trial and error.[4] Yet even considering the unlikely possibility that Menkaura did not have the same power and resources as

Khufu and Khafra, why should he build an 'inferior' pyramid at Giza to advertise this fact to posterity? There was plenty of land elsewhere. Menkaura's pyramid is by no means small, but the others are twice as tall and ten times as massive, reducing it to the status of dwarf. Why would he have settled for this?

One thing was certain: Menkaura knew that his pyramid was going to be much smaller than the other two at Giza. Such monuments have to be planned well in advance, and Menkaura must have approved the plan. Why approve a plan which would make him look inferior to his two predecessors? Whichever way we looked at it, there were flaws in the explanations. Perhaps we were looking at this pyramid, indeed any pyramid, in the wrong way. We were looking at each pyramid individually when we should have been looking at them as part of a unified project. All that we had to do was to think of the pyramids not as belonging to this or that pharaoh, but as a conglomerate of monuments devised as a unified plan. What was more likely is that the pharaonic state saw itself as custodian of the Memphite Necropolis as a whole, and that the chain of pyramids there were seen not as individual tombs but as an ensemble expressing the supreme ideologies of their rebirth cult. All the pyramids together made up the Necropolis or land of the dead; more accurately, they made up the Duat, the 'place where Osiris-Sahu is'. But how was the Necropolis linked to the stars of Osiris?

Returning to the Giza group, I saw that the questions had to be rephrased: Why did the master plan specify two large pyramids and one smaller? Why offset the smallest to the east? Now the answer became obvious: these 'anomalies' were not anomalies at all but constraints imposed in the planning, design and layout of a master plan which were reflected in the third pyramid. The next question was, therefore, were these constraints imposed by engineering or site problems or by religious considerations?

Trained in construction planning of layout, where the client's brief and the contours and area of the site, among other factors, imposed constraints on size and location of buildings, I knew by experience that many things which later appear as anomalies to others, are often planned aspects of the design. Even though no answer was as yet evident as to why the third pyramid was relatively small and offset from the south-west alignment of the two others, we could apply the process of strategic thinking in reverse: trace back what could have imposed those two criteria on the layout of the Giza pyramids.

There had been something else about the aerial photograph which now began to intrigue me, something important though it was not actually in the photograph: the River Nile. Not far to the east side of the Giza plateau was the lush Nile Valley and beyond it the city of Cairo. The river flows from the south to branch off just past Cairo into the wide Delta of Lower Egypt. The Nile's course, as French Egyptologist Jean-Philippe Lauer points out, 'flows quite exactly towards the north'.[5] In short, apart from the natural bends and kinks on its course, the river flow is meridional. Lauer also showed that all mastaba tombs from the First Dynasty onwards were orientated roughly south-north, parallel to the axis of the Nile. From the Fourth Dynasty, 'the orientation of pyramids reached a precision that was truly extraordinary'.[6] How, asked Lauer, did the ancient builders achieve such accurate south-north alignments? He believed the answer was that the ancients made use of stellar observations at the meridian transit of certain stars. Others before him, such as Edwards and the astronomer Zbynek Zäba, agreed with Lauer's hypothesis.[7] Zäba had also argued that the pyramid builders not only used stars for alignment but that they might also have been aware of precession.[8]

I knew that each pyramid at Giza was set so that the sides of its square faced a cardinal point. This meant that the monument was, intentionally or not, a fixed compass, easily directing due east, north, west or south depending on which side of the square base one stood. Despite this, the main axis of the pyramid ran along its meridian, especially looking from north to south. This is obvious because the entrance to the pyramid was always on the north face, so a visitor proceeded southward into it. The meridian was therefore the primary criterion for the original design and layout of the group. Yet here was another 'anomaly': the three pyramids of Giza, each set on a meridian, do not align together on a main meridian when seen as a group but are in alignment along their south-west axis, with the third pyramid offset to the east. What induced the architects to produce this odd layout?

The first factor to consider was the ground conditions of the Giza plateau, to see if the geology and contours of the site had forced this anomalous decision. But I knew the Giza plateau well, and there was nothing that would have prevented the planners from placing the three monuments in a row along a main meridian axis. Indeed, this would probably have been the easier choice.[9] Placing the three pyramids in a north-south axis would have meant two major quarries to the east and west of the project, which could have been used throughout the duration of the works. It would also, of course, have facilitated the alignment problem, with the need to set only one pyramid, the first, along a meridian. The alignment of the other two would merely have meant projecting the line farther south.

In the absence of major engineering constraints, there was really only one answer to the apparently illogical choice: the constraints or criteria which had determined the layout principle were based not on engineering logistics but on religious considerations. But what could these be? Most of the architect friends I consulted agreed on a symbolic rather than practical reason for the plan of the Giza group. They pointed out that most monuments – and especially intensely geometrical ones such as the pyramids – were charged with symbolic connotations. This often applied to the place where they were sited, its orientation and relative position to the geography of the area. In this case, the obvious geography and alignment to consider was the course of the Nile. The architects pointed out that the so-called Historical Axis of Paris, which extends from the Louvre to the new district of La Defense and goes through the Champs Elyseés, was orientated relative to the flow of the Seine adjacent to the Louvre.[10] Similarly in Washington DC, the main axis of Pennsylvania Avenue which the French architect, L'Enfant, had aligned, constituted another historical axis linking the White House with the Capitol; it, too, took account of the direction of flow of the 'sacred' Potomac River.[11] The pyramid builders of Memphis undoubtedly considered the meridional flow of the Nile when planning the Memphite Necropolis. But at Giza the general alignment of the three pyramids was not meridional but through a south-west axis.

Some explanation had to be found. The monuments were obviously of such importance to religious ideologies that any explanation had to correlate with the supreme belief in the rebirth of the kings who commissioned the project.

A common denominator was clearly at play here, yet it seemed to escape engineering logistics. Some other scientific discipline was required. The obvious astronomical layout of each pyramid, based on stellar observations, suggested that we should consider the alignments and layouts of the pyramids from the viewpoint of astronomy as well. I decided it was time to take a good long look at the stars.

6 GIZA AND THE BELT OF ORION

Man is a fallen god who remembers the Heavens
– Lamartine, *Meditations*

Let them be for signs and for seasons, and for days and years
– *Genesis i, 14*

Seek him that maketh the Pleides and Orion
– *Amos v, 8*

I The Rise of Orion

It was early November 1983 and, as is usual at that time of year, the night skies in central Saudi Arabia were remarkably clear. This was the time of week-end camping by expatriates in Riyadh in the golden dunes about twenty kilometres outside the sprawling western suburbs of this sedate city.

My wife, Michele, had packed the usual gear: alcohol-free beer, plenty of drinking water, food and the sleeping-bags. My daughter, Candice, was only four years old, but already a seasoned desert traveller. Two other couples with their children joined us. The idea was to select a high dune so that the kids could play on the clean, golden-coloured sand while the adults relaxed over hot coffee and an elaborate barbecue. We were all looking forward to escaping from the hard work and no play mood of Riyadh and the stifling atmosphere of a deeply Islamic society. Night on the dunes can be very beautiful. Immediately after the spectacular display of the setting sun came the darkness, with the canopy of a star-spangled sky almost at arm's length. Lying in my sleeping-bag, I counted the stars until I fell asleep.

For some reason I woke up at 3 a.m., perhaps subconsciously motivated. Once more I gazed up, at first unsure of where I was. High in the southern sky, arching over and almost marking for us the curve of the celestial equator, was a luminous band of light, resplendent against the inky black of space. It was the Milky Way and it looked like a great river in the sky. On its west 'bank' was a spatter of beautiful stars, brighter than all the others which surrounded them. I recognised them immediately as the constellation of Orion and went to wake up my friend Jean-Pierre, who shared my interest in astronomy and whose passion for sailing had necessitated his learning to navigate using the stars.

Silently, he came with me to the edge of the dune. Looking at the very bright star now high over the horizon, he let me into one of the secrets of astro-navigation. 'Do you know', he asked, 'how to find the rising point of Sirius once Orion has risen?' I shrugged my shoulders in ignorance. 'Well, first,' he said, pointing in the direction of the 'river bank', 'you must find the three stars of Orion's Belt. These three form a row and you extend the alignment downwards to the horizon. When the belt stars have risen about twenty degrees – roughly the height of an open hand at arm's length and with fingers outstretched – they will be followed by Sirius at the place on the horizon where they point.' He was now pointing towards the bright star on the horizon, which we both knew was Sirius. Then, almost as an afterthought, he uttered these words: 'Actually, the three stars of Orion's Belt are not perfectly aligned. If you look carefully you will see that the smallest of them, the one at the top, is slightly offset to the east and they are slanted in a south-westerly direction relative to the axis of the Milky Way. Also notice how . . .' At this point I cut him short. He gave me a puzzled look as I quoted the words I remembered only too well from the Pyramid Texts: 'The Duat has grasped the king's hand at the place where Orion is . . . [PT 1717]. O Osiris King . . . Betake yourself to the Waterway . . . may a stairway to the Duat be set for you at the place where Orion is . . . [PT 1717].' By now the others had woken up and joined us. 'Je tiens l'affaire!',[1] I cried excitedly. I had deliberately chosen the words uttered by Champollion when he realised he had decoded the secrets of Egyptian hieroglyphic writing and I hoped that someone in the group, a few of whom I had involved in the aerial photo puzzle of Giza, would catch on. From their expressions it was obvious they had not.

Jean-Pierre kept on looking intensely at Orion. 'What have you seen . . .?', he inquired, amused.

'The three pyramids of Giza', I said calmly.

'The what . . .?' asked Michele. She had heard endlessly about the star religion of the Egyptians in those last few months. 'Is this a joke . . .?'

'No, I am quite serious,' and I pointed to Orion's Belt. Thus began a saga which was to run for another ten years.

II Rostau: Gateway to the Stars

The idea that the Ancient Egyptian Duat, or heaven, had a counterpart on the land is something Egyptologists know from the many funerary texts extant in the museums of the world. The location, however, was always thought to be arbitrary, with no specific correlation intended. I knew that what I was suggesting was quite different. I read that in the New Kingdom the Duat, or rather its entrance, was thought to be at Abydos, then an important centre of Osirian worship. But I had also found out that in the Pyramid Age the Duat had its counterpart near Memphis, and that, in all periods, the Duat was said to have a central entrance or gate in a place named Rostau.[2]

I investigated further, and what I came up with confirmed what I had stumbled on that night in the desert. The way the three stars were slanted in relation to the axis of the Milky Way, the offset of the small star from the alignment of the two brighter ones, the southern shaft in Cheops's pyramid targeted to these very stars when the pyramid was built – all this was too much to be coincidence. Yet, if I was right, something as obvious as this had not only escaped the attention of Egyptologists but had probably done so because of the solar stamp given to the pyramids. It was not an easy consensus to break, so before rushing to proclaim my findings to Egyptologists, more research was required. The notion of Rostau was a starting point. If it could be shown that Rostau, the central gate of the Duat, correlated with the Giza necropolis, the central part of the Memphite-Duat, then we obviously had something. Naturally the deciding factor would be if other pyramids, especially those of the Fourth Dynasty, also correlated with other stars in the region of Orion. But first things first: where or what was Rostau?

Wallis-Budge, a former Keeper of Egyptian Antiquities at the British Museum and a prolific author, had made the startling comment that during the Pyramid Age the Memphite Necropolis containing the pyramid fields was known as the Duat of Sokar of Memphis. This god Sokar, a man with a falcon's head, was said to be the keeper of the Memphite Necropolis and, even more interestingly, was closely identified with Osiris during the Fourth Dynasty. This was confirmed by Dr Edwards who wrote that 'by pyramid times, Osiris had become identified with Sokar, the god of the Memphite necropolis . . .'[3] I also discovered that in many funerary texts the centre of the Duat was called Rostau. In the Shabaka Texts,[4] for example, the Memphite region is described: 'This is the land . . . [of] . . . the burial place of Osiris in the House of Sokar.'[5]

This prompted Selim Hassan to conclude that the centre of the Duat was not only identified with Rostau but with 'the kingdom of Osiris in the tomb'.[6] In the Book of the Two Ways, which contains funerary texts dating from the Middle

Kingdom period (*c.* 2000BC), we are also told that Rostau is the gateway to the necropolis and that it gives direct access to the Duat. The deceased tells us: 'I have passed on the roads of Rostau on water and land; these roads are those of Osiris; they are in the Sky . . .'[7]

Jane Sellers, who has for many years studied the astronomy of the Egyptians in relation to their texts, writes that 'the insistence in the Book of the Two Ways that the topography of the roads to Rostau, though in the sky, is on water and on land, hints at how the Egyptians conceived of the heavens'. She also suggests that 'the paths by way of water could have been the area which we know as the Milky Way'.[8]

Rostau is also mentioned in the Pyramid Texts in conjunction with the god Sokar (or Sokar-Osiris): 'For I am Sokar of Rostau, I am bound for the Place where dwells Sokar . . .' [PT 445]. The 'place where dwells Sokar' was, of course, the Memphite Necropolis, but it seemed also to have an astral location in the vicinity of the Milky Way. So was Rostau in the sky, that is Orion's Belt, to be correlated with the Giza pyramids?

So far there was good evidence that Rostau in the 'place where dwells Sokar' or Sokar-Osiris, was an actual place on land, somewhere in the Memphite Necropolis. This fitted the view of students of Egyptian symbolism, that it was 'vital to the spirit of Egyptian religion that the symbolism should be twofold', so that every affair of mankind was regarded as a 'repetition of some mythical happening in the time of the gods'.[9] The Egyptians believed that the gods, indeed the 'wisdom god' Thoth himself, had built the Giza pyramids during the golden age when gods lived on earth; the idea was later imparted to the Greeks, who also said that Hermes, the name they gave Thoth, had built the pyramids.[10] I remembered, too, that in the famous Westcar Papyrus of the New Kingdom the pyramid of Khufu, called the Horizon of Khufu, was linked to the sanctuary of Thoth, supposedly somewhere in Heliopolis.

Looking at a recently published *Atlas of Ancient Egypt*,[11] I was amazed to find that Rostau was near or indeed at Giza: a real place in the Memphite Necropolis, and the approximate location was given as 'southern Giza'. Indeed, Rundle Clark calls the god 'Sokar of Giza' seeing this place as being the ancient Rostau.[12] Many Egyptologists refer to Rostau as the ancient name of Giza. Goyon thought it was where the village of Giza is today,[13] and Rundle Clark says that 'Rosetau [sic] [is] . . . the modern Giza, the burial-place of Memphis and the home of a form of Osiris known as Sokar'.[14] Miriam Lichtheim, an eminent philologist at the University of California, says that Rostau was 'the necropolis of Giza'[15] and Faulkner similarly identified it with the 'necropolis of Giza or Memphis [and] later extended to mean the other world in general'.[16] In the Middle Kingdom and New Kingdom Osiris is called the 'august god in Rostau',[17] and it is implicit that Rostau was regarded as the place of great ritual where the reborn person can 'go forth into the day' as 'one who follows the god (Osiris) in his procession in his festival of Rostau . . . here begin the spells of the Fields of Offerings and spells for going forth into the day: of coming and going in the realm of the dead [Duat] . . .'[18]

It was clear that it could be argued that Rostau was not a mythical place but was indeed Giza, and that it was considered the gateway to the Duat region. What I now needed to confirm was whether the correlation I could see between the three Giza pyramids and the stars of Orion's Belt was part of a larger scheme.

III The Celestial River

As we have seen, the Pyramid Texts contain astronomical data in that they talk about observations made of Orion, of Sirius and other stars in the region of the sky the Egyptians called the Duat. What was thrilling and evocative was the way that the Ancient Egyptians correlated the Nile with the 'celestial river' i.e., the Milky Way, and this was known even by the Greeks. From the time of Homer, the Nile was associated with the mythical sky river called either Okeanos or Eridanus. The Hellenic historian, A. B. Cook, was of the opinion that

Eridanus (which today is a faint constellation formed by a string of stars joining Rigel to Alchermar) was 'at the outset none other than the Milky Way', and that in pre-Greek times, Okeanos 'simply meant the Galaxy' i.e. the Milky Way. Cook also drew attention to a statement by Hyginus that the river Eridanus was identified with the Nile, and that it was also often called Okeanos ('*Eridanus: hunc alii Nilum, complures etiam Oceanum esse dixerunt*').[19]

The identification of the Nile with Eridanus or Okeanos seems to have been common knowledge in the classical world. Even Diodorus reported that 'the Egyptians consider Okeanos to be their river Nile, on which their gods were born',[20] and the chronicler Eusebius says 'the Egyptians believe that the river Nile is the ocean from which the race of gods has taken birth'.[21] Much later Eridanus was identified with the River Po in Italy, and sometimes with the Rhine and even the Rhone, but as R. H. Allen remarks, 'none of these comparatively northern streams suit the stellar position of Eridanus, for it is a southern constellation, and it would seem that its earthly counterpart ought to be found in a corresponding quarter.'[22]

It is not hard to see why a Nilotic people with a sky religion should see a correlation between their river and the Milky Way. Just as the Nile divides Egypt into two regions, so the Milky Way divides the sky. It is quite probable that this relationship between the Nile and the Milky Way was what first gave the Nile dwellers the idea that a cosmic Egypt existed in that region of the sky which their souls could reach after their earthly existence. Wallis Budge explains this rather well:

> The Egyptians . . . from the earliest days . . . depicted to themselves a material heaven wherein Isles of the Blest were laved by the waters of the Nile . . . others again lived in imagination on the banks of the Heavenly Nile, whereon they built cities; and it seems as if the Egyptians never succeeded in conceiving a heaven without a Nile . . .[23]

Reading this, I was not surprised that the Pyramid Texts also tell us of an important 'Winding Waterway' in the eastern sky which closely resembles the Nile, with its own 'great flood' and 'fields' of reeds or rushes:

'May you lift me [the dead king] and raise me to the Winding Waterway, may you set me among the gods, the imperishable stars . . .' [PT 1759]

'Be firm, O king, on the underside of the sky with the Beautiful Star upon the Bend of the Winding Waterway . . .' [PT 2061]

'I have come to my waterways which are in the bank of the Flood of the Great Inundation, to the place of contentment . . . which is in the Horizon . . .' [PT 508]

'The Winding Waterway is flooded, the Fields of Rushes are filled with water, and I [the dead king] am ferried over thereon to yonder eastern side of the sky, to the Place where the gods fashioned me, where I was born new [reborn] and young . . . Lo, I stand up as a star which is on the underside of the sky . . . my sister is Sothis, my offspring is the Morning Star . . .' [PT 343–57]

It was now looking likely that I had stumbled upon the true mystery of the pyramids. The Duat, which stretched along the 'west bank' of the Milky Way corresponded to – indeed was seen as a mirror image of – that region we now call the Memphite Necropolis. It was, of course, not a necropolis at all in the Greek or western sense of the word; rather the Elysian Fields, the earthly counterpart of the heavenly abode of the king-gods of Egypt – the Egypt, that is, of the Pyramid Age.

IV Development of the Orion Correlation Theory

The evidence was now mounting that the Ancient Egyptians viewed the area of the Memphite Necropolis as a terrestrial image of the heavenly Duat. Throughout antiquity the Milky Way was looked upon as a celestial river analogous to the Nile and in Giza we had, quite literally, Orion's Belt on the ground. What I now needed to check was what the Pyramid Texts had to say concerning the pyramids, not as metaphors for a religious idea or symbol but as material structures. It was then that I discovered something very curious: the Pyramid Texts make few direct statements concerning the pyramids themselves, and these are all huddled together in one long passage, known as Utterance 600.

In this Utterance Ra, the sun god, offered his benevolent protection to the monument in question. As head of the Heliopolitan pantheon and ancestral father of the gods, including Osiris, this was not unexpected, much as we might ask for the protection of God the Father while believing in our resurrection through Jesus Christ. Ra, the sun god, might indeed protect the pyramid and the whole Necropolis, but it was through Osiris that rebirth was deemed to be achieved. Finally in Utterance 600 I found what I was looking for: an unequivocal statement that connected the king and his pyramid construction to Osiris. The statement was an instruction to his son, the new Horus-king, to proceed to the pyramid fields: 'O Horus, this (departed) king is Osiris, this Pyramid of his is Osiris, this construction of his is Osiris, betake yourself to it . . .' [PT 1657].

To understand this better, we should remember that versions of the Pyramid Texts have been found in not one but several pyramids.[24] It therefore makes sense to suppose that this Utterance is meant not only for one specific king but serves as a general liturgy for all departed kings. In the plural Utterance 600 reads: 'O Horus, these (departed) kings are Osiris, these Pyramids of theirs are Osiris, these constructions of theirs are Osiris, betake yourself to them . . .' [PT 1657].

I at last understood that we were being told, in plain language, that the pyramid constructions were to be considered Osiris. As I already knew that the celestial form of Osiris was Sahu, and that this figure corresponded with our modern constellation of Orion, the pyramids were indeed Orion too. The text writers could not have made their intent plainer or more straightforward, and it substantiated my theory that the three pyramids of Giza were symbols of Orion's Belt.

My next step was to find further visual evidence. I had a good photograph of the three stars of Orion's Belt and was able to place it against the aerial shot of the three Giza pyramids. The correlation was stunning. Not only did the layout of the pyramids match the stars with uncanny precision but the intensity of the stars, shown by their apparent size, corresponded with the Giza group: there were three stars, three pyramids, three Osiris-Orion kings.

As I read the word Osiris I began to conjure the sky image of Orion, the 'soul of Osiris'. Utterance 600 was dealing with an afterlife ritual, not so much with the embalmed corpses of dead kings but with their souls, and more specifically their astral souls which joined Osiris-Orion in the celestial Duat. Osiris in this case was, of course, also Osiris-Orion. Thus, the passage would read: 'O Horus, these (star souls of departed) kings are Orion-Osiris, these pyramids of theirs are Orion-Osiris, these constructions of theirs are Orion-Osiris, betake yourself to them . . .'

Suddenly I realised that not only the three Giza pyramids but others too might have stellar positions in the Memphite Necropolis. Now that the Giza group identified with Orion's Belt, it could be used as a reference or datum point from which the relative positions of other stars of the Duat could be located. The two great pyramids of Dashour for example, and those which had been located at Abu Ruwash and Zawyat Al Aryan which flank Giza, might not these also correlate to stars of the Duat? Surely all Fourth Dynasty pyramids would have been involved in the master plan to forge the soul of Osiris on the sacred land of Memphis? I recalled excitedly that two of the pyramids in question, those of Djedefra at Abu Ruwash and Nebka at Zawyat Al Aryan, bore star names: 'Djedefra is a Sehetu Star' and 'Nebka is a Star'.[25] A 'Sehetu Star' meant a star of the Duat. What star might that be? The temptation to investigate further was compelling.

I laid out a map of the Memphis area and compared it with a picture of the region of the sky containing Orion. Carefully aligning the Giza group pyramids with the stars of Orion's Belt, I saw that the pyramid of Djedefra at Abu Ruwash corresponded with the star Saiph or Orion's 'left foot' and that at Zawyat al Aryan represented Bellatrix in his 'right shoulder'. There were no known pyramids in locations to match other stars such as Betelgeuse and Rigel, so I could only conclude that these had never been built or that they had long since been demolished and had disappeared under the sands of the Western desert. Given the ruined state of the pyramids of Zawyat al Aryan and Abu Ruwash, this is not an unlikely supposition. Five of the seven bright stars of Orion were thus accounted for in the Fourth Dynasty pyramids.

The pyramids of Dashour, however, posed a problem. They were not part of 'our' modern Orion figure, and it was only much later that I worked out where they fitted. What was clear at this stage was that what we now call 'the Orion correlation theory' had generated a momentum of its own.

It now seemed like the right time to approach the experts and see what they thought about it.

THE STAR CORRELATION THEORY 7

I think you have made a very convincing case . . .
– I. E. S. Edwards, Keeper of Egyptian Antiquities at the British Museum (1954–1974), letter to author, October 1984

In my opinion your theory is not capable of independent verification . . .
– T. G. H. James, Keeper of Egyptian Antiquities (1974–1984), letter to author, December 1983

I The Experts Speak

Late in 1983 I prepared a brief paper with a few hand sketches and posted the Orion Correlation Theory, as I now called it, to the British Museum. I was still living in Riyadh and I knew how notoriously slow the mail was to Europe. The reply came much quicker than I thought. It was a letter from Professor T. G. H. James, then Keeper of Egyptian Antiquities. This position had previously been occupied by Dr I. E. S. Edwards from 1954 to 1974, and many other eminent names, such as Sir Wallis Budge and Samuel Birch, had held it. Dr James's reply left me nonplussed: he told me that while he thought that my theory fitted some of the facts, it would be difficult to accept it as an explanation for the construction and placing of the Giza pyramids. He pointed out that the theory could not be applied to the two pyramids at Dashour and maintained that there is no real evidence from antiquity to support it.

I was disappointed by the apparent lack of enthusiasm. I agreed with him that many questions still needed to be answered, such as the matter of the two pyramids of Sneferu at Dashour, but I was taken aback by his seeming dismissal of the theory. I wondered what would constitute 'independent verification' and why he thought that 'there is no good evidence from antiquity to support' my theory? Were not the statements in the Pyramid Texts, the Badawy articles on the shafts in Cheops's pyramid that pointed to Orion's Belt, and now the layout plan of Giza, 'good evidence'? At least sufficiently compelling to warrant a closer look at the theory? I had obviously not struck the right chords, and could only assume that Dr James's letter was a tactful way of saying that the correlation between the three Giza pyramids and the three stars of Orion's Belt was no more than coincidence.

My experience had taught me that collections of coincidences do not occur easily. Coicidence is a word we all use when we cannot explain why there is a convergence of certain events and facts. What is coincidence to some, is not so to those who understand the links between the events and the facts. The facts before us were not remote or detached from one another. The Pyramid Texts, compiled in the Fifth Dynasty, were surely expounding events witnessed during the Fourth Dynasty, which immediately preceded the compilation of the Pyramid Texts. These, as we have seen, told us in no uncertain terms that the departed Osiris-king became a star in the constellation of Osiris-Orion. Then there was the shaft in the Cheops pyramid which Badawy and Trimble agreed pointed to Orion's Belt when the pyramid was built. There was also the anomalous size and offset of Menkaura's pyramid, which could only be explained by a correlation plan with Orion's Belt.

All this – and there would be more – was 'good evidence' to me, especially when we were trying to solve a mystery more than 4400 years old. Indeed, considering the remoteness of the event, we were lucky to have any shred of evidence at all.

In September 1984 I took a short holiday in England. As soon as I arrived in London, I decided to pay a visit to the British Museum, meet Dr James and see what else could be done to persuade him to take the matter up seriously. Dr James, however, was not available. A young assistant, I think it was Dr Carol Andrews, was very helpful and when she saw that the subject matter concerned the pyramids, she advised me that it would be better handled by Dr Edwards, the previous Keeper of Egyptian Antiquities. Although he had retired in 1974, he was still very active in the field, and was currently the vice-president of the Egyptian Exploration Society. There was no question that Edwards was seen by most scholars as the supreme authority on the subject of Egyptian pyramids, and his views would not only be more valuable, but would carry more weight. It was agreed that I should send the relevant papers as soon as possible, which would then be forwarded to Dr Edwards. I posted these from France a week later. His reply arrived in Riyadh in October 1984, and the views he expressed certainly differed from those of his successor. The letter is reproduced with his kind permission:[1]

> 16 October 1984
> Dear Mr Bauval,
>
> Thank you for your letter dated 8th September, which reached me after being posted in France last week.
>
> Let me say that I found your astronomical observations very interesting, and I think you will see from the enclosed article, which I wrote for the Dows Dunham *Festschrift* four years ago,[2] that I am very much in agreement with your contention that the stars in Orion's Belt were an important element in the orientation of the Great Pyramid. I think you have made out a very convincing case that the other two pyramids at Giza were also influenced by it. I

> have sent a new edition of my book (*The Pyramids of Egypt*) on the pyramids to the publishers (Viking Press and Penguin Books) and it is about to go out to the printer. According to present expectations, it will be out next summer and it will embody the substance of the enclosed article.

Dr Edwards then entered a brief commentary on my ideas related to the measurements of the Great Pyramid, feeling that, with such a geometric shape, a mathematician could make it fit any number of different measurements. He then gave his own conclusions about the stellar connotations which I had revealed to him:

> The position of Osiris in the Fourth Dynasty is still very uncertain. Since the earliest Pyramid Texts date from the end of the Fifth Dynasty they do not provide much evidential help.
>
> In your contention that the pyramids are intended to represent stars I wonder whether the truth is not that the pyramids were intended to enable the king to reach the stars. In my view this was the purpose of the step-pyramids and the true pyramids, which generally embodied a step-pyramid, were intended to enable the king to reach both the solar and astral heavens.
>
> I am,
> Yours truly,
> I. E. S. Edwards

Though we differed on some interpretations, Dr Edwards's view that I had presented a convincing case was very encouraging and much appreciated at this stage. I was beginning to feel isolated, and it was good to discover that an authority as eminent as Edwards was highly in agreement with my contention that the stars in Orion's Belt had played a major role in the orientation of the Great Pyramid and its two companions at Giza.

A few months later, in January 1985, I received a letter from Dr Jaromir Malek, director of the Griffith Institute of Oxford University at the Ashmolean Museum. Dr Malek surprised me by saying that he had 'no special astronomical or mathematical knowledge and the few comments I can make are of a purely Egyptological nature'.

> . . . I wholeheartedly agree with you that astronomical observations and mathematical calculations played an important part in the design, construction, and perhaps even siting of Egyptian pyramids . . . [and] . . . in the paper itself, I would be prepared to consider seriously the observation that the Giza pyramids were positioned or sited in a manner as to represent the three stars of Orion.[3]

He also commented on the 'civil calendar' of Ancient Egypt, and felt that my 'putative date' for its introduction was incorrect. These items, though, were not an important aspect of my theory and had been the subject of academic debate for many decades. Dr Malek then said, in relation to my suggestion that other pyramids should be investigated in the light of a stellar correlation siting: 'I also fully agree that the other groups of pyramids would have to be examined bearing this in mind, and this, in my opinion, is the only path one can take to make some progress in the matter.'

Dr Malek made a final comment on the stellar correlation theory: 'To write that "the ancient Egyptians saw the land of Egypt as being an 'image' of the sky" is overstating the case. To base further theories on it is unsafe, to say the least.'

I had now to let this be. For the next year I was totally occupied with more pressing concerns related to my 'real' work and my personal life. The company I was working with was starting a new project in Saudi Arabia and there was much to keep me fully occupied. Also, my wife Michele and I were planning to settle in Australia after our long stay in Saudi Arabia. A new member of the family had arrived in December 1984, our son Jonathan, and it was time to look for a more congenial place to raise the children. Sydney was where the rest of my family had gone after our mini-exodus from Egypt in 1967, and it seemed the logical place to consider.

I did go to England in November 1985 and met Dr Edwards in his home near Oxford. A most charming and affable man in his late seventies, Dr Edwards was on his way to London, but we had a brief conversation on the new stellar ideas in the Pyramid Texts and the pyramids. Edwards was of the opinion that scholars had neglected these Texts for more exciting subjects, and agreed that the stellar element in the Texts had been ignored. However, he reiterated his view that the true pyramids were solar symbols, and though they might have retained some stellar notions in their design, the influence, he believed, was predominantly solar. I said politely that I begged to differ. He smiled and recalled that he did not know where I came from. Alexandria, I replied. 'Ah, I somehow thought so,' he said. 'A place where new ideas often came from . . .'. He said that should I want to publish my ideas one day, he might offer some suggestions on the matter, and I assured him that I would take him up on that. I did; two years later. In the years to come we became good friends, and though we differed on the symbolic interpretation of the true pyramids, it did not prevent us from contributing to each other's views on the pyramids and allowed us to share many happy moments with Rudolf Gantenbrink after he made his historic discovery in 1993. But all this was still a long way off.

Michele, the two children and I arrived in Australia in September 1986. We bought a house in Sydney's northern suburbs near my sister's home, and settled into the gentle pace of suburban living. I decided to work part-time and return as well to the issue of the pyramids. I found, to my delight, that the Mitchell Library of the University of Sydney was well stocked with Egyptological books. Many professional journals were regularly received and outsiders, like myself, were free to make use of the library as guests of the university. I was to spend many long hours devouring all I could about the Egyptian pyramids, astronomy and religion. I consulted hundreds of books and articles and my photocopying bill was enormous. Yet once launched, I could not be stopped. I bought a second-hand computer and began the big adventure of putting my findings and theories into article form. I was not sure where or when they might be published, if at all. But I was sure of one thing: it was my responsibility, and I had to get it off my chest.

While in Australia I made the acquaintance of Dr John O'Byrne, Professor of Astronomy at the University of Sydney. He offered to do the necessary precessional calculations for me and to verify my astronomical commentaries. His calculations confirmed the accuracy of the Badawy-Trimble discovery. The southern shaft of the King's Chamber, taken to slope 44.5 degrees, had pointed to Orion's Belt in *c.* 2600BC. There was an odd discrepancy, though, which puzzled me. The calculations showed that the shaft was aimed more specifically at the central star, Al Nilam (Epsilon Orionis), than at Al Nitak (Zeta Orionis) which, according to the Giza–Orion's Belt correlation, was the star which should correspond to the Great Pyramid. Since the precessional motion was now in its upward cycle,[4] I also asked Dr O'Byrne to try for me the slightly later date of *c.* 2500BC. This brought the shaft target closer to Zeta Orionis, but not exactly on it. Either the date needed refining, or the slope Petrie had given needed to be verified. It was then that I remembered the southern shaft of the Queen's Chamber. Dr O'Byrne had shown me where to find the 'rigorous formula' in the standard Catalogue 2000.0 to calculate precession, and had said that a good pocket scientific calculator would be adequate to get values within the arc minute level. I bought myself the most powerful one on the market: a Casio fx-8000G which could memorise the precessional formula.[5]

I took Petrie's value of 38 degrees 28 minutes for the slope of the southern shaft of the Queen's Chamber and looked at a sky map with Orion on the southern meridian. It had to be a star below Orion's Belt. But which? I looked again. Sirius, the star of Isis! Why had I not thought of this before? Then I remembered: the shaft was supposed to be abandoned. Why bother with an abandoned shaft? I supposed that this was what Badawy and Trimble had thought. Well, it took only a few minutes on the calculator, so why not try? I chose a date of *c.* 2650BC, a little earlier than that of *c.* 2600BC which tallied with the other southern shaft higher up the pyramid. I reasoned that the lower shaft would have been started decades before, so *c.* 2650BC was a fair estimate. After the adjustment for the proper motions of Sirius, which are quite considerable (see Appendix 1), I got a declination of −21 degrees 20 minutes. Working the altitude for the position of Giza, I got 38 degrees 41 minutes, almost spot on to the 38 degrees 28 minutes slope given by Petrie. We now had the two southern shafts pointing respectively to Osiris-Orion and Isis-Sirius for the epoch *c.* 2650–2600BC. Coincidence surely had to be ruled out. I wondered what the Egyptologists would make of this now.

I was, however, still troubled by a niggling discrepancy: the southern shaft of the King's Chamber should really have pointed precisely at Al Nitak (Zeta Orionis), the lowest star of Orion's Belt and not the central one. The correlation was too accurate, and so was the astronomical alignment of the base of the pyramid and its slopes, to make it likely that on such an important astronomical matter the builders would have 'missed' the specific star corresponding to this specific pyramid, even if it was by only half a degree of arc.

I worked out the precession for Al Nitak again, this time allocating it the altitude of 44.5 degrees and working back the epoch. This gave me the date of *c.* 2590BC. Then I worked out the epoch for Sirius at altitude 38 degrees 28 minutes and got *c.* 2730BC. This meant an unrealistically long period of 140 years between the two shafts; so much time could surely not have elapsed between the start of the Queen's Chamber and that of the King's Chamber. About twenty years was the limit that I (and others, I was sure) considered acceptable. Something was wrong either with Petrie's values or with the way the shafts had been constructed; the former seemed more likely, in view of the precision of the work elsewhere in the Great Pyramid. To get a date reading that made good sense, various trials with the scientific calculator indicated that the slopes should be slightly steeper for both southern shafts, with the Queen's at nearer 39.5 degrees and the King's nearer forty-five. Only then would a date of twenty years separating them be obtained.

This would give a dating of *c.* 2450BC for the Great Pyramid, a century or so 'younger' than hitherto assumed. Could that be possible? And could Petrie's measurements be slightly out? No one would have the answer until Rudolf Gantenbrink measured the angles again in 1993.

While preparing my articles, I decided to probe a little more with the Egyptologists, this time in the United States. I sent a brief dissertation to the University of California at Berkeley, and in August 1986 received a reply from Dr Frank A. Norick, Principal of the Lowie Museum. Dr Norick admitted that he and his colleague, James Deetz, were 'fascinated with some of [my] correlations and conclusions'. They did not feel they were in a position to evaluate the thesis and had passed it on to Professor Cathleen Keller of the Department of Near Eastern Studies. In her reply, she said that she would rather wait for the 'work of Mr Mark Lehner', who was conducting a topographical survey of the Giza plateau, to be completed, but this is what she felt about my theory at present: her opinion was that while there was ample evidence in the Pyramid Texts to connect the dead king with Orion she did not feel that the layout of the Giza monuments was predetermined by the Orion constellation. She then provided confirmation of an inherent problem in Egyptology which I was beginning to suspect existed widely in the profession: namely that the serious study of the connection between astronomical phenomena and Ancient Egyptian architecture is in its infancy and that it was taking a different form from what she termed 'the (often wild) conjectures of "pyramidiots"'.

There followed her warning, which I have quoted earlier, that the association of celestial bodies with architecture made Egyptologists uncomfortable and '. . . more afraid that connections *do* exist between the orientation . . . of Egyptian temples and the heavens, than that they do not'. But what was telling was her comment about 'pyramidiots'; this was the core of the problem. Mention a 'theory' on the pyramids, especially one that involves the stars, and Egyptologists shy away. Circulating my theory through the international circuit of Egyptologists was not getting me anywhere; the star correlation stood little chance of surfacing in those areas. The best I could hope for was encouragement from others, such as Dr Edwards.

It was high time to publish. Yet where and how? Another trip to England seemed in order. I was determined to take up Dr Edwards's offer to recommend me to the editor of an Egyptological journal.

II A Forum: Discussions in Egyptology

In England, I rented a car and drove to the little village north of Oxford where the Edwardses had their home, and where Dr Edwards and I discussed yet again our favourite subject. Engaged in pyramid discussions, Dr Edwards radiates an enthusiasm which is refreshingly stimulating and his openness to all viewpoints and new ideas is very appealing.

He told me of a new Egyptological journal run by his friend, Dr Alessandra Nibbi, which was open to non-Egyptologists with a contribution to make. The journal was called *Discussions in Egyptology*. I liked the name; it had an open feeling about it. It seemed that an article had appeared not long ago by the engineer, John Legon, who had made a good case that the Giza group of pyramids was part of a unified plan, though Legon's approach was entirely mathematical, with no mention of the Pyramid Texts or stellar ideas.[6] Dr Edwards promised to recommend a contribution from me to Dr Nibbi. The next day I telephoned her and she offered to take two articles, provided, of course, that they were of the style and seriousness expected by her readers. I assured her they would be, and said I would send them to her, with the accompanying photographs and diagrams, as soon as I returned to Sydney. I sent them early in June 1988, and in July Dr Nibbi told me that the articles would appear in volumes 13 and 14 of *Discussions in Egyptology* (*DE*).

Michele and I had meanwhile taken the decision to relocate to England. We left Australia in May 1989 and found a house halfway between London and Oxford. The kids went to the

1a The authors in front of the Giza Pyramids

1b Overhead view of the Giza group

2a The Step Pyramid of Zoser at Saqqara

2b The Fifth Dynasty Pyramids at Abusir

3a The 'Bent' Pyramid at Dashour

3b The Red Pyramid at Dashour

4a Statue of Mariette outside Cairo Museum

4b Maspero the discoverer of the Pyramid Texts

5a Burial chamber in the Pyramid of Unas showing the Pyramid Texts

5b Pyramid Texts with group of three stars from Unas Pyramid

5c Pyramid Texts that say Unas–Osiris

6 ***The Giza overhead***

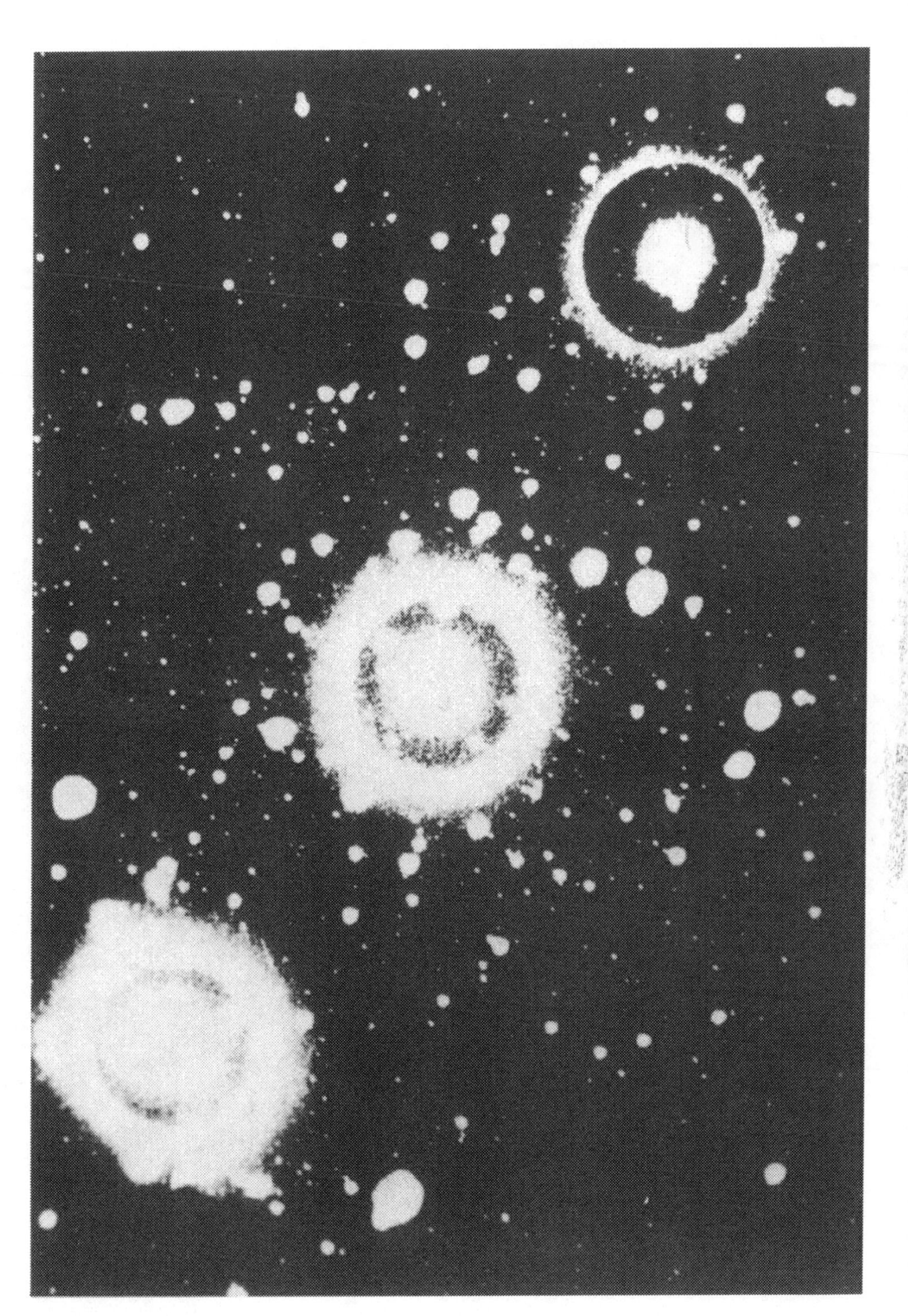

7 *The stars of* Orion's *Belt*

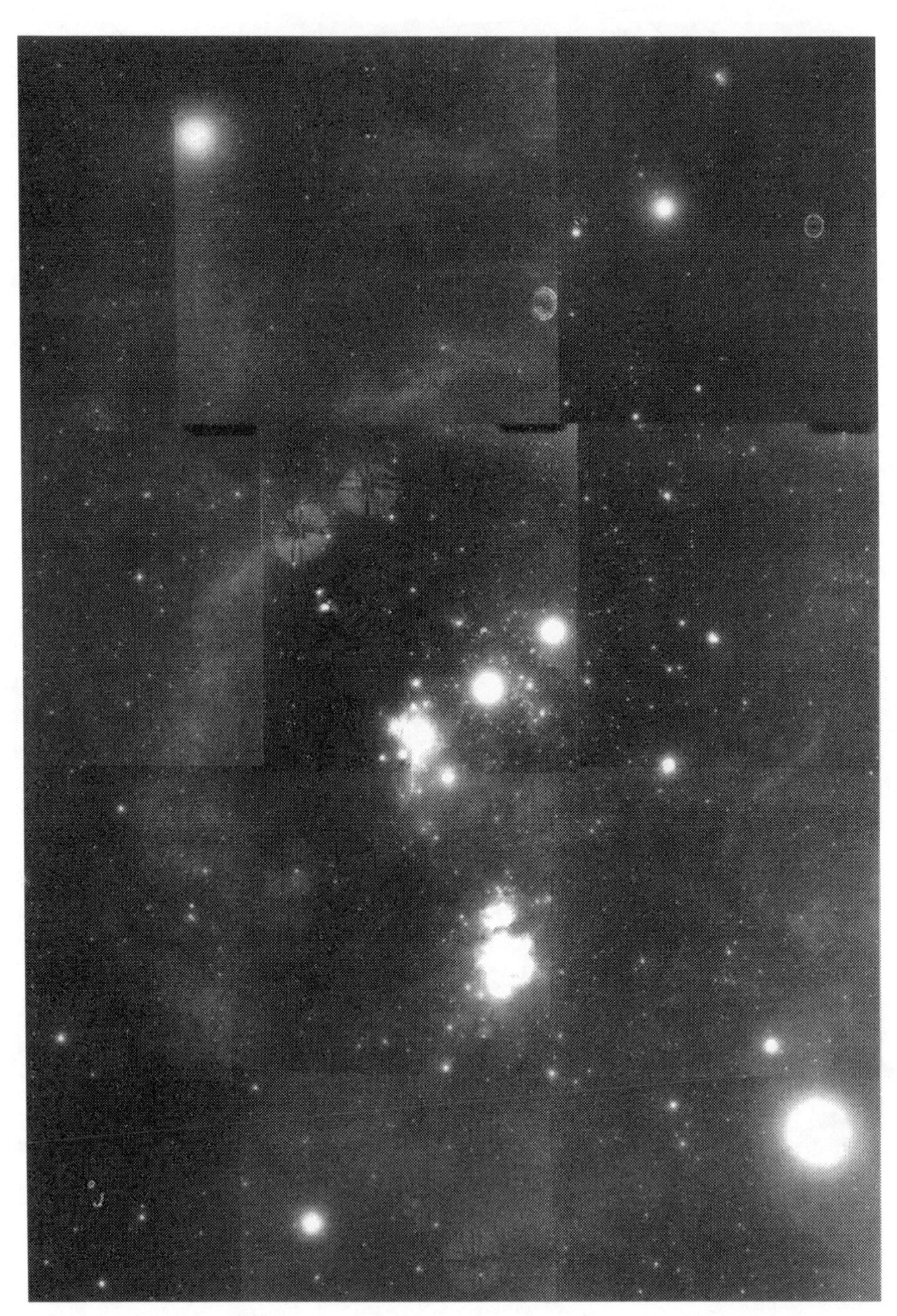

8 ***The constellation of Orion***

9a The niche in the East wall of the Queen's chamber

9b The empty sarcophagus in the King's chamber

10*a* ***The authors in front of the southern shaft of the Queen's Chamber days before the discovery of the 'door' by*** UPUAUT 2

10*b* UPUAUT 2

11*a* UPUAUT 1 *going up the southern shaft of the King's Chamber*

11*b* *Iron plate found in the southern shaft of the King's Chamber in 1837 by* R.J. Hill

12a Robert Bauval and Rudolf Gantenbrink in Munich

12b Dr I.E.S. Edwards and Robert Bauval on 6 April 1993 after the showing of the UPUAUT *video*

13a **The Benben Stone from the pyramid of Amenemhet III in the Cairo Museum**

13b **The Sahu-Orion figure on the Benben of Ahmenemhet III**

14a Oriented iron meteorite 'Willamette' in the Smithsonian Institute, New York

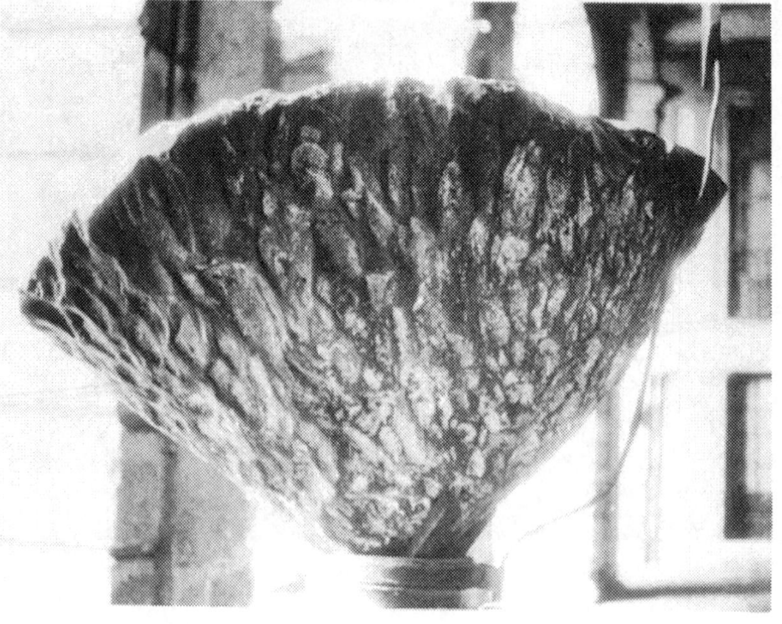

14b Oriented iron meteorite 'Morito' in the Institute of Metallurgy Mexico City

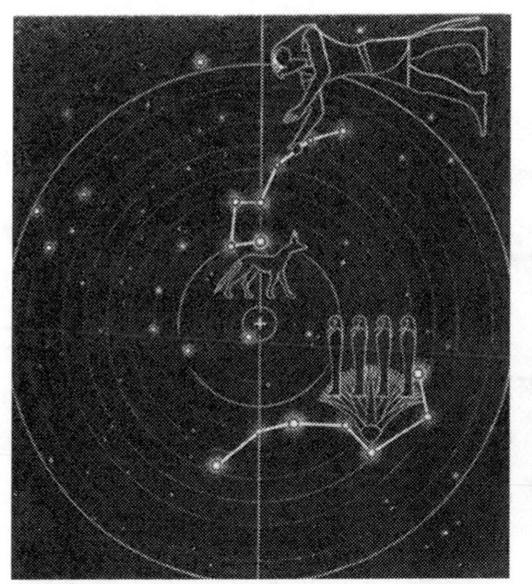

15a (left) Horus holding the 'adze of Upuaut' (Ursa Minor) aligns with the northern shaft of the Queen's Chamber, and, simultaneously, Orion-Osiris (right) rises in the east

15b The Opening of the Mouth Ceremony depicted in the Papyrus Ani in the British Museum

16 ***Artist's impression of the stellar landscape, showing Osiris (Orion) and the shaft of the Great Pyramid pointing to his belt***

local school and I, too, went back to study. I had decided that a postgraduate degree in European business and marketing would come in handy in a united Europe. In the excitement of making a new home in England and the activities of the postgraduate course, I almost forgot about my articles. Then in May a large parcel was delivered by the postman: three complimentary copies of *Discussions in Egyptology* volume 13.

The Orion Correlation Theory was at last officially published, nearly six years after I had made the fateful observation in the Saudi Arabian desert. The article in *DE* 13 was entitled 'A Master Plan for the Three Pyramids of Giza Based on the Configuration of the Three Stars of the Belt of Orion', and included six pages of text, four photographs and two diagrams. It was written in academic style, lacking the excitement I really felt, sticking to facts and evidence, and heavily cross-referenced and annotated. I made no mention of pyramids other than the Giza group, and avoided a discussion on the shafts in the Queen's Chamber. This was for later.

The second article in *DE* 14, was entitled 'Investigation on the Benben Stone: was it an Iron Meteorite?'. In it I discussed the sacred relic of Heliopolis in terms of its stellar and Osirian connotations (see Chapters 11 to 13). Finally, in January 1990, Dr Nibbi accepted a third article to complete the stellar thesis, with the title 'The Seeding of the Star Gods: a Fertility Rite Inside Cheops's Pyramid?' This article revealed the Isis-Sirius target of the southern shaft of the Queen's Chamber, and put seriously into question the established consensus that it had been abandoned. It also contained my thoughts on the position of the openings of the shafts. Knowing that an important aspect of the rituals had been the flaunting of ithyphallic statues which symbolised the king's potency and fertility, and that a fertility ritual was described in stellar terms in the Pyramid Texts, involving Isis-Sirius and Osiris-Orion with the mention of a stellar phallus (the Belt of Orion shafts?), I began to see evidence of an extraordinary fertility ritual inside the Cheops pyramid, in which the shafts played a major role. Their role was not simply in the sending of the pharaoh's soul to the phallic region of Osiris-Orion (the Belt stars) but is for the symbolic seeding of a Horus-king. The relevant passage in the Pyramid Texts addresses Osiris-Orion:

> Your sister (wife), Isis, comes to you rejoicing for love of you. You have placed her on your phallus (shaft?) and your seed issues into her, she being ready as Sothis (Sirius), and Horus-Sopd (a star) has come forth from you as 'Horus who is in Sothis' [PT 632]

The article pointed out that a similar fertility ritual involving the king and a high priestess was known to have taken place in ancient Mesopotamia in a chamber inside the stepped-pyramid ziggurats.[7] This ritual involved the 'Morning Star', seen as the great cosmic goddess Ishtar apparently identified to the planet Venus, and commemorated the New Year (Akitu) and the fertility that the flooding of the Euphrates brought to the land. In parallel, the Egyptians celebrated the New Year with the heliacal rising, the annual flooding of the Nile; Sirius being the great cosmic goddess Isis (incidentally identified later with Ishtar) and also involved a 'Morning Star'. The tentative conclusion was that 'the contents of this present article should compel us to suppose that a fertility ritual not unlike the one performed in the ziggurats of Mesopotamia may also have been performed inside Cheops's pyramid and possibly in other pyramids as well'.[8]

Little did I suspect then that in March 1993 Rudolf Gantenbrink would prove that the Queen's Chamber and its shafts had not been abandoned as Egyptologists said but, on the contrary, may have been the most important ritualistic elements of the whole pyramid cult. Never in my wildest dreams would I have suspected that in 1990 the Isis-Sirius shaft would make the front pages and the news in a dozen international newspapers.[9]

In the mild spring of 1990 I deluded myself that my mission was over. I had got the theory into print and Egyptologists, astronomers and other scholars could make what they wanted of the new stellar findings. It was as if a heavy burden had been removed from my shoulders; an original idea which involves public interest was a cumbersome load to cart around. There had been many times when I felt a strange anxiety, a disquieting feeling that I would not get through, and that the Orion–Giza correlation would be lost again in a timeless zone. I was thrilled and relieved that it was over, but I also felt a curious sense of loss. I would miss the excitement of research and even those long, lonely hours in libraries, but I told myself firmly that the personal quest was over.

So, in March 1990, as one unsympathetic Egyptologist had advised me when I began my quest, I resolved to 'abandon this subject and try to become a good engineer'. I took up freelance consulting and tried to persuade myself that the pyramids were best left to the Egyptologists. But each time I looked up at the sky and saw the stars of Orion, I wondered about those silent monuments in Egypt and could almost feel their frustration at not being understood. Try as I might I was unable to abandon the subject altogether: for one thing there was still the question of the Dashour pyramids and how they fitted into the plan. It was only a matter of time before I would be drawn back full-time into the Orion Mystery.

8 THE BROTHER OF OSIRIS

Seth . . . originally connected with the Hyades, the V-shaped, head-like part of our constellation, TAURUS. As the brother of Osiris, his position in the sky was adjacent to ORION . . . an important court decision gave the office of Osiris to Horus, and Seth was banished to a position bearing the 'southern' constellation of ORION . . .'

– Jane B. Sellers, *The Death of Gods in Ancient Egypt*

I The Southern Pyramid Fields of Dashour

With the awareness that a correlation or duality existed between the sky-Duat and the Memphis-Duat on the ground, and that the central region was expressed by the Giza-Orion's Belt correlation, I had a sort of map of the Duat of Memphis. Although the evidence so far was compelling, I also knew if the theory was to hold water, not just the three Giza pyramids but the other four pyramids of the Fourth Dynasty had to be considered in the stellar correlation of the Duat of Memphis. These were the two large pyramids of Sneferu at Dashour in the southern part of the Memphite Necropolis, and those allocated to Nebka and Djedefra at Zawyat Al Aryan and Abu Ruwash.

In my first article in *Discussions in Egyptology*, I had left the issue open by asking, 'does this master plan include a wider correlation between the geomorphy of the sky landscape about Orion and the landscape about the Giza Necropolis?'[1] Now, in 1992, this question had to be answered.

I had long been aware that two other Fourth Dynasty pyramid sites – Zawyat Al Aryan and Abu Ruwash – flanked the three Giza pyramids in much the same way as the stars Saiph and Bellatrix in Orion flanked the three stars of Orion's Belt. These pyramids, as we have seen, also had star names: one, 'Djedefra is a Sehetu (Duat) Star' and the other 'Nebka is a Star'.[2] The stars in question had to be those of Osiris-Orion in the Duat, the stellar destiny reserved for these kings. It all fitted neatly together, with most of the pattern of Orion – five of its seven main stars – correlated to Fourth Dynasty pyramids in the Memphis-Duat. With 'Bellatrix' located south-east of Giza, it was not difficult to see how the three or four little stars forming Orion's 'head' could fit the three or four[3] little pyramids at Abusir, a kilometre or so south-east of Zawyat Al Aryan. Indeed, in the Westcar Papyrus, which speaks of Khufu (Cheops) and his 'horizon' (his cosmic pyramid), specific mention is made of the 'three children' of a priestess of Heliopolis who were said to have founded the Fifth Dynasty and who erected their small pyramids at Abusir.[4] A fifth pyramid, now lost, is also believed to have been built at Saqqara.[5] Whoever inscribed the Pyramid Texts in Unas's pyramid, and whoever commissioned them, were living at the close of the Fifth Dynasty, and had on display in the Memphis Necropolis all that was already built there. As far as the true geometrical pyramids are concerned, these included all those of the Fourth Dynasty plus the three or four smaller pyramids at Abusir. In stellar terms of the Osirian Duat, these made up the 'leg' (Abu Ruwash=Saiph), 'phallus' (Giza=Orion's Belt), and 'shoulder' (Zawyat Al Aryan=Bellatrix) of the giant Osiris-Orion. Yet one of his most evocative features was his fully extended 'arm', seen on many drawings such as the pyramidion of Amenemhet III, with the open hand cupping a bright star. In Greek mythology this star is Aldebaran in the Hyades, and marks the position of the mace of Orion the Hunter or the Giant. Their angular distance placed the Hyades roughly in the correlation map of the Memphis-Duat with the position of Dashour, which demarcated the southern portion. I now had a pretty good idea where and what to look for.

The pattern of the stellar Duat was defined by the cluster of stars, from the Hyades to Canis Major with Orion inbetween, all found on the 'west bank' of the Milky Way. On land, in the Memphis-Duat, this corresponded to the pyramid fields from Dashour to Abu Ruwash, with Giza somewhere in the middle, all found on the west bank of the Nile. With the three Giza pyramids sited in ancient Rostau and fitting the location of Orion's Belt in the centre of the sky-Duat, the implication that Dashour was to be correlated to the Hyades was indicated by the layout principle of the master plan. The Giza group had shown that the layout was based on the heliacal risings of the stars of the Duat and their projection to the ground, each represented with a pyramidal monument fixed on a meridian. The things to look at, then, were the heliacal rising of the Hyades for *c.* 2550BC, the estimated time of Sneferu's reign, and the meridians of each of his allocated pyramids at Dashour. If this hunch was right, a relationship between the two should be found.

In February 1992, just before sunset, I was on a British Midland flight making its landing approach to Cairo. The approach was from the west, the plane flying low over the Memphite Necropolis, and the view below was breathtaking. I could see all of the major pyramids, from Giza to Dashour, their west faces catching the orange light of the setting sun. As the plane crossed over Giza, the two pyramids of Dashour revealed their meridional layouts, that of the northern one offset to the west of the southern pyramid. Like Giza, this was another 'anomaly' which, again like Giza, would be elucidated by a stellar siting.

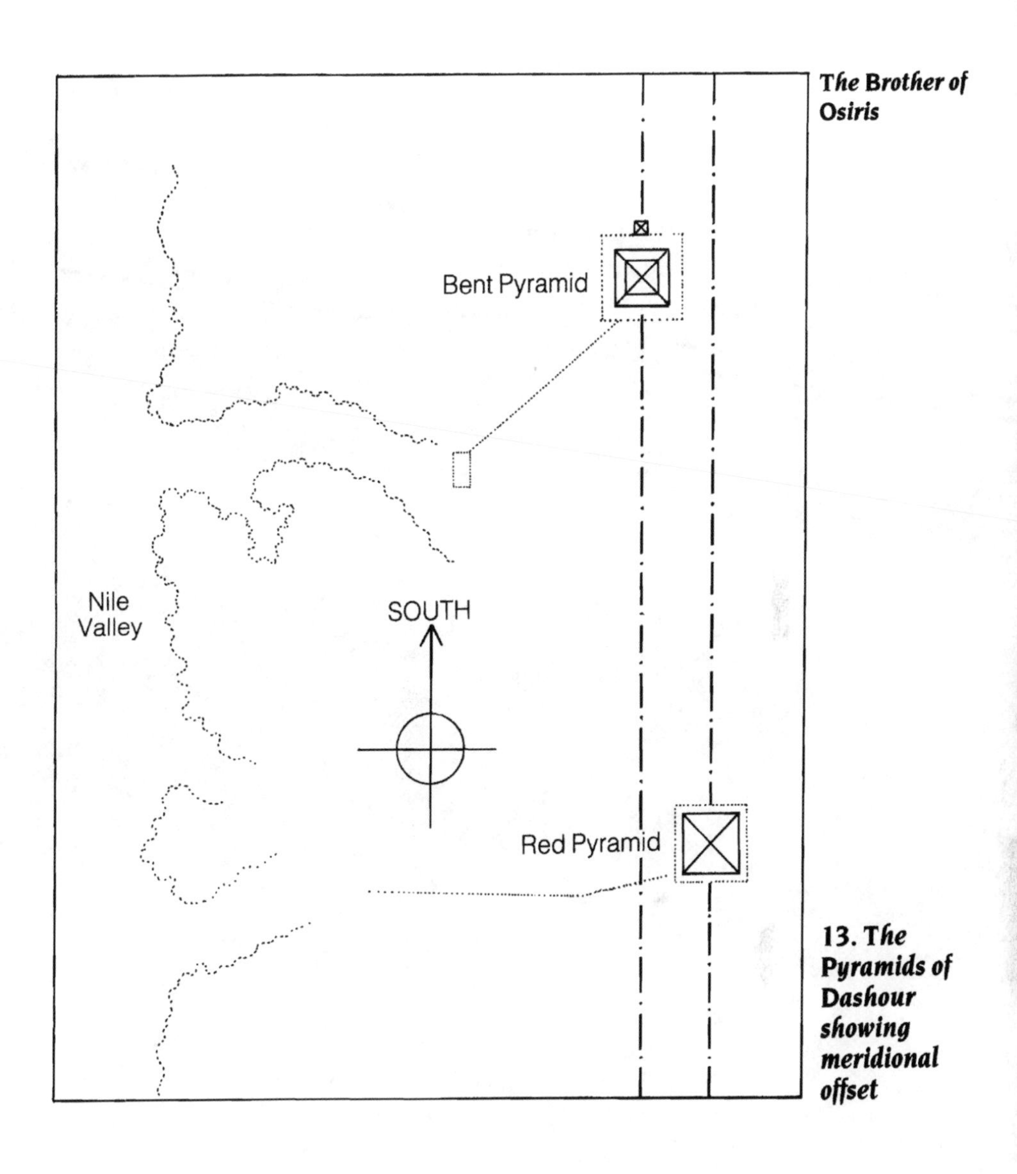

13. *The Pyramids of Dashour showing meridional offset*

The ancient architects again presented us with a curious discrepancy. After setting the axis of one pyramid along a meridian, they set the axis of the second pyramid some 500 metres to the west and the pyramid itself about 1850 metres farther north.[6] This was an odd choice. It would have saved them many problems if the same meridian had been used, and would have been better to have the two sites closer so that preliminaries, such as labour huts, open quarries, ramps from the Nile and other organisational requirements, could be deployed for both sites with the works on each pyramid staggered to the required rate of progress.[7] Again, this choice of different meridians and the pyramids 1850 metres from each other contradicted engineering logistics. It therefore had to be religious and, as at Giza, this meant astronomical. Recalling that the anomalous offset of Menkaura's pyramid had been

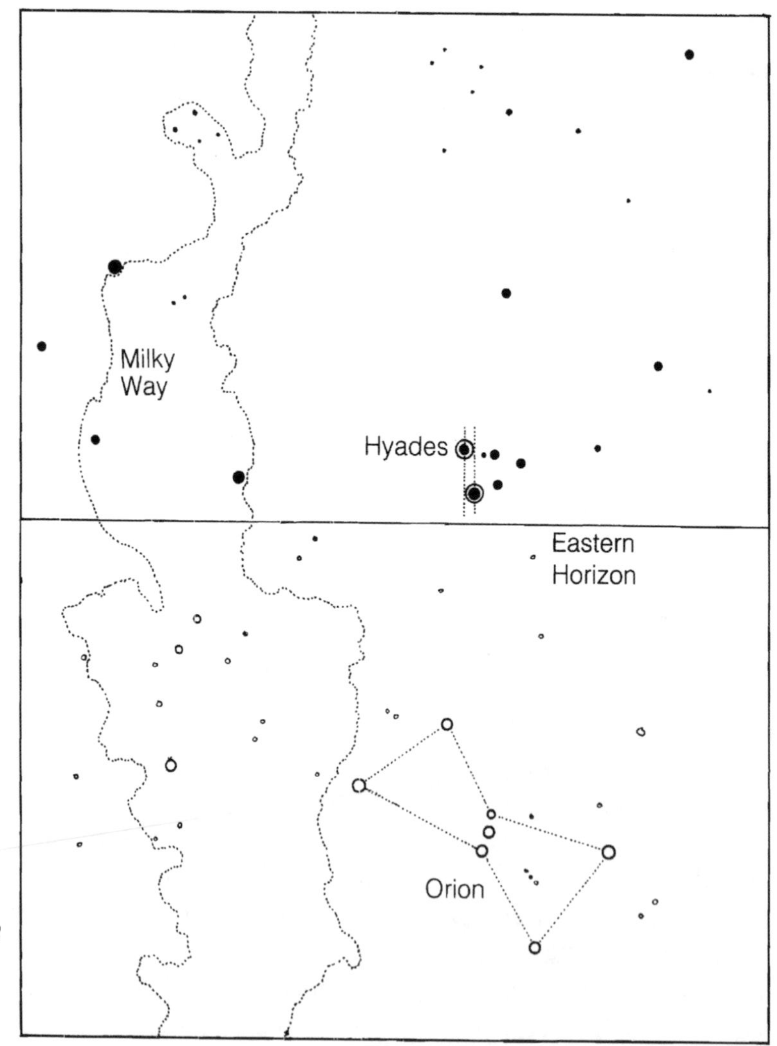

14. The Rising of the Hyades

imposed by the configuration of the stars of Orion's Belt, and the south-west alignment by the slant of the three stars relative to the axis of the Milky Way, a similar situation ought to have prevailed for the Hyades in the sky and Dashour on the ground.

Back in England, I again used the Skyglobe 3.5 computer program to simulate the rise of the Hyades when the reign of Sneferu began. My reasoning was that, if Khufu's reign began in *c*. 2450BC,[8] the reign of his father would have begun in about 2475BC, since Sneferu was believed to have reigned for some thirty-four years.[9] Adjusting the epoch on the computer for 2475BC, I found that the Hyades would rise heliacally during April. Looking at the odd triangular or V-shape of this ancient constellation, the two stars at the base of the triangle were Aldebaran and another catalogued No. 311 (Epsilon Taurus). No. 311 rises first, due east, and when it is at three degrees altitude Aldebaran follows. These two stars, seen together after rising, had the exact layout relative to each other and the axis of the Milky Way as the two Dashour pyramids relative to each other and the axis of the Nile. Transposing the two stars on the correlation Memphis-Duat map, they fitted the position of the two Dashour pyramids. This gave us the complete stellar pattern of the sky-Duat, from the outstretched hand of Osiris (the Hyades) to his leg (Saiph) and took into account all the Fourth Dynasty pyramids and the cluster of small Fifth Dynasty pyramids at Abusir. It was as though the mist lifted to show a new landscape, clear and sharp in the image of Osiris-Sahu.

At about this time, while I was looking for the meaning of the Hyades in Egyptian astronomy, I came across a recent book which confirmed that I was not alone in exploring the influences of stellar precession on the Ancient Egyptians. Jane B. Sellers, an American Egyptologist whom we have mentioned earlier, had done an elaborate study on this subject and presented it in *The Death of Gods in Ancient Egypt*.[10] What really got me interested in her work was that the Hyades and Orion were very much in her mind too. Using the powerful Lodestar V.202 astronomical computer program, she had come up with some startling findings. These, coupled with her knowledge of ancient Egyptian religion and texts, made her work exactly what I was hoping to find.

II The Lady of Precession

Jane B. Sellers is described by her publishers as 'having spent much of her sixty years questioning puzzles in the fields of astronomy and ancient Near Eastern civilisations'. She comes across as one of those grand ladies of the stamp of Maria Reiche, the German mathematician who has devoted her life to studying the Nasca lines of Peru near the Andes highlands.

After getting a degree at Goddard College in Vermont, Sellers studied Egyptology at Chicago's Oriental Institute. A keen admirer of the late Dr Giorgio de Santillana, historian and author of *Hamlet's Mill*,[11] she has broken new ground for modern Egyptology by drawing attention to the need to use astronomy and, more particularly, precession of the stars for the proper study of Ancient Egypt and its religion. Her main focus, like mine, has been on the Pyramid Texts and the so-called Memphite Theology. With the aid of scientific astronomy, she has sought to explain the development of the religious ideas of pre-dynastic and early dynastic Egypt. In her words:

> Archaeologists, by and large, lack an understanding of the precession, and this affects their conclusions concerning ancient myths, ancient gods and ancient temple alignments. Philologists, too, ignore the accusation that certain problems are not going to be solved as long as they imagine that familiarity with grammar replaces scientific knowledge of astronomy. For astronomers, precession is well-established fact; those working in the field of ancient man have a responsibility to attain an understanding of it.[12]

A proper and detailed review of Sellers's thesis is outside the scope of this book. Briefly, one of her important contentions[13] is that the Ancient Egyptians had noticed the precessional changes of the stars even though they may not have understood them scientifically. She also feels that they had even worked out the rate of change, and brings a range of arguments to support her views. In all this I fully agreed with her. Precession had to be taken into account if the basis of the religious rituals of an ancient people was to observe the stars diligently. In Appendix 2 we have provided a full discussion on precession, but a short paragraph to explain its effects is in order before proceeding with our discussion.

Precession is caused by a very slow motion of the earth, a sort of wobble that takes about 26,000 years to complete a full cycle. The effect is not real but apparent, and only involves the stars. The stars do not actually move but appear to move because of the earth's precessional wobble. To show the effect, take Orion's Belt as seen from Giza. Imagine it sitting on the meridian, due south. Today it is at 59 degrees altitude above the south horizon. In the time of the Pyramid Age, *c.* 2500BC, it was much lower, at about 45 degrees. In about 10400BC it was even lower, at 11 degrees. The precessional effect is also clearly visible at the rising of Orion's Belt in the east: imagine Orion's Belt just over the eastern horizon at rising time. Today it rises almost due east, at azimuth 91 degrees. *C.* 2500BC, it rose farther to the south of east, at azimuth 106 degrees. In 10400BC, even farther south of east, at azimuth 169 degrees. The full cycle of precession, if we measure the effect at the meridian, consists of a half-cycle of 13,000 years from, say, maximum to minimum altitude, and another half-cycle of 13,000 years from minimum to maximum altitude. For Orion's Belt, since about 10400BC a half-cycle has started at minimum altitude of 11 degrees above the horizon (observed from Giza). It then slowly moved in an upward direction so that by the Pyramid Age it was at 45 degrees above the horizon, and today it is at 59 degrees.

This acts as a sort of star-clock for our planet. Knowing the exact rate of precessional change and the co-ordinate of a star,[14] we can determine its altitude, say, at the meridian for any given epoch or, if you prefer, its rising point on the eastern horizon. We deduced that the southern shaft of the King's Chamber pointed to the star Zeta of Orion's Belt, so precessional calculations give us, with a fair degree of accuracy, the period of *c.* 2450BC.

Returning to Sellers's thesis on the astronomy of Ancient Egypt, her principal premises for fixing certain prehistoric and early historic events rested on her belief that the Ancient Nile dwellers not only noticed precession but focused their attention on the spring equinox. There, I did not agree with her. It is not clear what importance the spring equinox had for Ancient Egyptians, other than the sun's reaching mid-point in its annual changes, the effect noticed at rising or setting, or at the meridian transit. But this also applies for the autumn equinox. Those who have studied Egyptian religion or astronomy agree that the period of the year which dominated the mind of the early Nile dwellers was, without a doubt, the summer solstice. In the epoch immediately preceding the Pyramid Age, the summer solstice coincided with the heliacal rising of Sirius and the start of the Nile's flood, and it was on this fascinating conjunction that many of the cultic ideas were based. The heliacal rising of Sirius also denoted the New Year and served as the basis of calendrical computations; Egyptologists and archaeo-astronomers are at one on this. E. C. Krupp, the well-known archaeo-astronomer and director of the Griffith Observatory in Los Angeles, wrote:

> The Nile, with its annual flooding, made civilisation possible in Egypt . . . [it] was the real ruler of Egypt . . . The apparent connection between celestial and terrestrial phenomena greatly affected the Egyptian view of the world . . . [they] considered the heliacal rising of Sirius to be so important that they marked the beginning of the new year by this event. Even more compelling was the fact that the heliacally rising Sirius and the rising Nile coincided, approximately, with the summer solstice . . .[15]

The astronomer James Cornell was of the same opinion:

> From the very time the first humans settled in the Nile Valley, the periodic event of prime importance to their lives – their very survival – was the annual

> flooding of the river . . . this cyclical event, crucial to the establishment of Egyptian civilisation, also led naturally to the concept of time . . . [and] the development of the calendar. . . . By happy coincidence . . . Sirius first appeared in the morning sky around the summer solstice and at about the same time as the start of the Nile flood . . . the length of the Egyptian solar year was thereby set as the interval between the successive (heliacal) risings of that star.[16]

It is crucial in the study of Ancient Egyptian religion in relation to astronomy that the importance of the summer solstice be clearly appreciated. Not only did it mark the apogee of the sun's annual changes in declination but it provided a rough marker for the 'year' and, more importantly, the coming of the annual flood. This last was the real 'mystery' of Egypt: the fact that the waters began to rise at the time of the summer solstice and the heliacal rising of Sirius deeply affected the psyche of the Ancient Egyptians. A sort of stellar mechanical omen was witnessed in the sky a few weeks before the Sirius-summer-solstice-flood conjunction, and this, of course, was the appearance at dawn of fully risen Osiris-Orion.

Sellers's greatest contribution, in my view, was to bring home the fact that without the tool of scientific astronomy and a basic knowledge of observational astronomy and precession, it is impossible to interpret correctly the mass of funerary texts and rituals and (I might add on her behalf) religious monuments. In that, she has rendered a great service to Egyptology. Another important point which she raised, and one which was to enlighten me in my Dashour-Hyades correlation, was her conclusion that Seth, the brother of Osiris, was from very early times identified with the Hyades.[17] She also drew my attention to the astronomical connotations in the Memphite Theology, a theological tract based on the legandary unification of Egypt,[18] often also called the Shabaka Texts. In 1987 I had suspected a strong astronomical value in these texts but had shelved studying it until later.[19] Sellers's comments not only regenerated their importance but triggered the answer to the

Dashour-Hyades correlation and tied up a lose end in the Memphis-Duat correlation.

III A Black Stone

There is a slab of stone, a single block of black granite measuring about 1.3 metres by 1.5 metres, in the British Museum, classed as Item No. 498. On it are carved dozens of lines of hieroglyphic inscriptions, many unfortunately damaged when this stone was used in modern times for grinding wheat.[20] Some call them the Shabaka Texts; to others they are the Memphite Theology. Although the stone dates from the Twenty-fifth Dynasty (*c.* 710BC) the inscriptions are believed to be copies from sources as far back as the Pyramid Age. The American philologist, Miriam Lichtheim deduced that the language used in the inscriptions on the Shabaka stone 'much resembles that of the Pyramid Texts' and takes this as evidence of ancient sources.[21] This view is held by many scholars, including Jane Sellers.[22]

The Shabaka Texts begin with a curious introduction by the scribe commissioned to copy the texts. It seems that the pharaoh, Shabaka, wanted to have preserved for posterity certain ancient writings which were worm-eaten (presumably written on papyrus or wood) and ordered that they be copied on to a black granite slab, the Shabaka stone. In the ancient scribe's words: 'This writing was copied out anew by his majesty in the house of his father . . . for his majesty found it to be a work of the ancestors which was worm-eaten . . .' [1–2].

The first part of the text seems to be a sort of Solomon's judgement on the apportioning of the 'Two Lands' (Egypt) between Seth and Horus after the death of Osiris. Let us briefly reiterate the story: Seth was the brother of Osiris and Horus was the son of Osiris. The story begins with Seth and Horus being called by Geb, the earth god. Geb was the legitimate husband of the sky goddess, Nut, mother of Osiris and Seth. As such, he was the legitimate father of Osiris and, by virtue of his earth role, the highest authority on territorial matters.[23] After some deliberation,

> He made Seth king of Upper Egypt . . . [and] made Horus the king of Lower Egypt up to the place where his father (Osiris) was drowned, which is at the 'division of the Two Lands'. Thus Horus stood over one region and Seth over one region. They made peace over the Two Lands at Ayan. That was the division of the Two Lands. [7–9]

This allusion to Osiris being 'drowned' is another version of his death, not at the hands of Seth, but by drowning in the Nile somewhere near Memphis. Ayan is thought to be outside the north wall of the city of Memphis, and seems to have marked a frontier line which separated the kingdom of Seth (Upper Egypt) from that of Horus (Lower Egypt). At the time the text was written, this would have implied a line of demarcation dividing the Memphite Necropolis just south of Zawyat Al Aryan and, by necessity, would have created a lower or northern part of the Duat of Memphis containing Orion proper (Giza), and an upper or southern part containing the Hyades (Dashour).

Immediately after taking this seemingly fair decision, Geb had second thoughts and retracted it. His new decision was to give both kingdoms to Horus. This, of course, created a major conflict between Horus and Seth, and an epic battle ensued, with Horus being the victor. Horus was thus the 'uniter of the Two Lands' and was so acclaimed in the Memphite Theology: 'He is Horus who arose as king of Upper and Lower Egypt, who united the Two Lands in the Nome of the Wall (Memphis), the place in which the Two Lands were united. [13c–14c].

The Texts then inform us of the true meaning and particular sanctity of this holy place adjacent to Memphis:

> This is the Land . . . [of] the burial of Osiris in the House of Sokar (Memphite Necropolis) . . . [you must call] Isis and Nephthys without delay, for Osiris has drowned in his water . . . Horus speaks to

> Isis and Nephthys: 'Hurry, seize him . . .' Isis and Nephthys speak to Osiris: 'We come, we take you . . .' They brought him to [the land]. He entered the hidden portals in the glory of the lords of eternity . . . thus Osiris came into the earth at the royal fortress, the north of [the land] to which he had come . . . There was built the royal fortress . . . [17c–22]

Isis urges Horus and Seth to 'fraternise so as to cease quarrelling in whatever place they might be'. [15c].

As with the Pyramid Texts, I gave the Shabaka Texts the chance to speak for themselves. They provided an image of the body of Osiris lying along the west bank of the Nile, stretching over the demarcation line between the southern and northern part of the House of Sokar (the Memphite Necropolis). The story has a cosmic ring and suggests the same imagery in the sky-Duat. In this 'place', which is 'at the north of the royal fortress' (obviously Memphis) we are told that 'Horus stood over one region and Seth stood over one region . . . that was the 'division of the Two Lands'.

There has long been speculation why the dividing line or border between Lower and Upper Egypt was made at Memphis. The usual suggestion that it was an ideal strategic location is not really tenable; in later times Memphis stopped being the seat of pharaohs and the capital was transferred to Thebes, nearly 1000 kilometres upstream. It should be remembered that Egypt is an elongated country, a 1200-kilometre stretch formed by the narrow Nile Valley. A demarcation line of the 'two lands' at Ayan near Memphis would divide Egypt rather unevenly, with Lower Egypt running northwards only 220 kilometres to the Mediterranean coast, albeit with the rich Delta region, whereas Upper Egypt would stretch 1000 kilometres from Memphis to Aswan, hardly a fair parcelling of Egypt for two feuding pretenders.

I began to get the impression that the 'land' in question was not all of Egypt as we know it, but a holy region with a cosmic duality and which specifically contained the 'House of Sokar' (the Memphite Necropolis). In the Memphite Theology we are not dealing with a typical territorial dispute but with a cosmic event, with the protagonists, Horus and Seth, considered as 'gods'. After the mythical death of Osiris, the real prize to be shared was the god's earthly domain, that is the earth-Duat of Memphis, which now contained the 'Pyramid Fields', the symbols of pharaonic theocracy and the material expression of the state religion.

In cosmic terms this 'land of Osiris-Sokar' was the starry Duat along the west bank of the Milky Way/Celestial Nile; there, too, Osiris was lying along a region contained by Canis Major in the lower sky and the Hyades in the upper sky with the constellation of Orion between them. But what, I wondered, was the 'border' that supposedly divided the lower sky and the upper sky? Was there some feature separating the Hyades from the rest of the starry Duat of Osiris?

I considered the location of Ayan immediately north of Memphis and traced the demarcation line as a latitude going through the Memphite Necropolis. The line passed just south of the Abusir pyramid field, with Saqqara and Dashour at its south (Upper Egypt), and Abusir, Zawyat Al Aryan, Giza and Abu Ruwash at its north (Lower Egypt). The original decision by Geb would have thus given Horus the lower portion of the Memphis-Duat containing the pyramids of Abusir, Zawyat Al Aryan, Giza and Abu Ruwash, and the upper portion to Seth, containing the pyramids of Saqqara and Dashour.

The body of Osiris in the sky was the giant Sahu sky-image, which we saw as a striding man with one arm outstretched, the open palm cupping a star. The sexual part, i.e., the phallus, is clearly fixed with the stars of Orion's Belt, and these must have evoked the sexual potency and seeding of the stellar Osiris. In the Osiris-Isis myth, the crucial moment was the making by Isis of an artificial phallus so that she could fertilise her womb with the seed of Osiris, and become pregnant. Oddly, there exists an ancient text called the Inventory Stela which is in the Egyptian Antiquities Museum in Cairo.[24] Its date remains a mystery, though Egyptologists date it at around 1500BC. It was found by Mariette in 1800, while he was excavating the ruins of a small chapel called 'The House of Isis', next to Cheops's pyramid. This stela refers to Cheops and the Great

Pyramid and nominates Isis as 'Mistress of the Pyramid'.[25] If this is correct, it is tempting to visualise the artificial phallus as being the southern shaft of the King's Chamber in Cheops's pyramid, which was aimed at Orion's Belt, the phallus region of the Osiris-Sahu image in the sky. This reminds us of that passage in the Pyramid Texts concerning the stellar copulation and seeding ritual between Osiris and Isis:

> Your sister (wife), Isis, comes to you rejoicing for love of you. You have placed her on your phallus (shaft?) and your seed issues into her, she being ready as Sothis (Sirius), and Horus-Sopd (a star) has come forth from you as 'Horus who is in Sothis' [PT 632]

What now needed to be deciphered from the Shabaka Texts was why Geb had given the southern portion of the Memphis-Duat to Seth, only to revoke his decision soon afterwards. Was there an event in the sky-Duat which might have made Geb consider that the corresponding Memphis-Duat had to be separated into an upper and lower portion? Was there a change in the position of the Hyades, for example, which moved it from the lower sky into the upper sky *c.* 2500BC? And what feature divided the sky into an upper and a lower part?

IV The Equator in the Sky

Seen from the earth, the sky appears as a huge hemispherical vault covering the flat and apparently circular land, its bottom rim resting on the horizon. In scientific astronomy we separate the east and west sides of this celestial hemisphere by the meridian, an imaginary line conjured as running overhead from due north to due south. We also separate the celestial hemisphere into a south and a north side, with an imaginary line, the celestial equator, running due east to due west, but the line is directly overhead only if you are on the earth's equator; otherwise it always inclines towards the south,[26] crossing the meridian line at an altitude equal to ninety degrees *less* the latitude where you are standing. Thus, if you are near London, the celestial equator crosses the meridian at 90 − 51 = 39 degrees altitude above the southern horizon. If you are near Cairo, it crosses the meridian at 90 − 30 = 60 degrees altitude over the southern horizon. The celestial equator is, therefore, the astronomical 'border' which divides the upper and lower skies.

Jane Sellers was to conclude that

> Seth . . . originally connected with the Hyades, the V-shaped, head-like part of our constellation, TAURUS. As the brother of Osiris, his position in the sky was adjacent to ORION . . . an important court decision gave the office of Osiris to Horus, and Seth was banished to a position bearing the 'southern' constellation ORION . . .[27]

The gigantic Sahu sky-figure stretched from the Hyades (the 'southern' constellation of Sahu) past Orion proper and finally to Canis Major and Sirius. How did the celestial equator divide this 'land'?

Running Skyglobe 3.5, I went to epoch 3100BC, when Egyptologists say the unification of the two lands of Egypt took place, and then projected the sky-Duat star region (Hyades, Orion and Canis major) on the meridian. The celestial equator passed just above the Hyades, meaning that they were in the lower sky (corresponding to 'Lower Egypt'). Knowing that the precessional effect caused an upwards shift of the stars, I decided to see when the Hyades, and especially the two stars, Aldebaran and 311 (Epsilon Taurus), which I equated to the two Dashour pyramids, would cross the celestial equator and move into the upper sky ('Upper Egypt'). I went up the centuries, 3100BC, 3000BC, 2900BC UNTIL 2000BC. I was astounded to see on the monitor screen the events of the unification as explained in the Memphite Theology. It was a thrilling sight! Bearing in mind that the celestial equator is at zero declination, and that negative declinations are in the lower sky and positive declinations in the upper sky, I refined the dates to the nearest decade and the readings obtained are reflected in the table.

Epoch	Declination	
	Aldebaran	Star 311
3100BC	$-5° 35'$	$-3° 29'$
2450BC	$-1° 56'$	zero
2080BC	zero	$+2° 13'$

These precession events were very revealing; they showed that at exactly the time King Khufu (Cheops), the alleged builder of the Great Pyramid, came to power, star 311 was poised to cross the celestial equator and leave the lower sky for the upper sky. Then, in *c.* 2080BC, at the time when the Pyramid Texts were put into the Fifth and Sixth Dynasty pyramids, the same happened to Aldebaran. In correlation, this meant that the Dashour (Hyades) pyramids now 'belonged' to Upper Egypt, a territorial dispute settled not by land deeds but by the precessional mystery of the stars. No priest could confront the decision of the office of the sky gods, the Great Ennead of Heliopolis.

The basis of archaeological and chronological evidence leading Egyptologists to date the unification of the Two Lands at *c.* 3100BC was therefore not confirmed by precession; this suggested that it was at a later date, possibly after 2400BC, and thus *after* the Fourth Dynasty, not before.

Precession does not depend on archaeological or chronological interpretations; it relies on the natural cyclical period of the precessional wobble, and thus behaves as the true epoch marker, a great star-clock behaving according to the laws of natural physics. I now began to see that the unification was prompted by the shifting further north of a sacred demarcation line or divider latitude, an event not to be considered as the start of dynastic Egypt but a religious dispute that occurred after the Fourth Dynasty. Such a dispute, though evidence was scant, was suspected by many Egyptologists including Dr Edwards, who indicated a political upheaval at the close of the Fourth Dynasty by noting that '. . . although documentary records are lacking, the character of the political events which attended the close of the Fourth Dynasty may be conjured from a number of indications'.[28]

These indications, according to Edwards, are the adoption of the suffix ra in the royal names: Kharf-ra, Menkau-ra, Djedef-ra, Sahu-ra and so on. To him this meant that the solar cult was gaining authority and becoming the state cult, because of the incorporation of the sun god's name into the royal names. Also, the term 'Son of Ra' became part of the title of pharaohs 'from the Fifth Dynasty onward', even though the Horus-name, denoting the king as Son of Osiris remained the dominant title of kings.[29] There is too, of course, the material evidence of the drastic decline of pyramid construction: Fifth and Sixth Dynasty pyramids became smaller and their masonry was of much poorer quality, also an indication of political upheaval or cultic change.

The Memphite Theology seems, therefore, to narrate in mythological and cosmic terms a real dispute over the throne of Egypt which occurred at the close of the Fourth Dynasty. If that is so, the golden age of Osiris's reign came to an end with the completion of the Giza necropolis, and a dispute ensued over who should inherit his pharaonic legacy. The title 'son of Ra' may have been used by a pretender who claimed direct descent from the head of the 'father' of the Heliopolitan pantheon of gods, the Great Ennead, to gain supremacy over any pretender claiming to be the son of Osiris.

Oddly enough, this seems to be confirmed by the Westcar Papyrus.[30] In 1947, Edwards drew attention to this mysterious document, which reveals the story of the coming to power of the first three kings of the Fifth Dynasty, as seen by the Ancient Egyptians themselves.[31] Two of these, Sahura and Neferirkara, built their pyramids at Abusir.

V The 'Triplets' of a Priestess

The Westcar Papyrus has preserved an ancient legend concerning the creation of the Fifth Dynasty which, we are told, came about when a high priestess of Heliopolis was seeded by Ra, the sun god.[32] This was a typical ploy used when a dynastic change or coup was in the making. For example Olympias, the mother of Alexander the Great, claimed that Zeus-Ammon had made love to her, making her son the ultimate contender to the throne of Macedonia and Greece;[33] Caesar claimed a descent from Venus, generatrix.[34] Divine intervention in matters of dynastic disputes was an easy way to sway the credulous populace to believe a shaky or even illegitimate claim to the throne. A 'miraculous' birth was always effective and as late as the seventeenth century in Europe they were still going strong. Louis XIV of France for example was said to have been conceived miraculously,[35] when after twenty-six years of sterile union between Louis XIII and Anne of Austria, the couple produced a 'solar' heir, nicknamed Dieudonné, (God-given).[36]

The claim of a solar pregnancy by the priestess of Heliopolis was probably a carefully orchestrated plot and seems to have worked. According to the Westcar Papyrus, Ra came down to earth and seeded the all-too-willing wife of the high priest at Heliopolis. This resulted in her giving birth to triplets, all of whom were to become kings of Egypt: the pharaohs Userkaf, Sahura and Neferirkara. With the Westcar Papyrus, as with the Memphite Theology, I believe we are dealing with a historical event explained in cosmic terms, which resulted in the creation of the Fifth Dynasty. The site chosen for their pyramids by this new solar dynasty, to express its connection with the Fourth Dynasty and its dominant astral cult of Osiris, was the flat region of Abusir. Here a triplet of little pyramids was built which, in the Memphis-Duat correlation, denoted the 'head' of Osiris-Sahu sky-image. I believe there is an astronomical event which connects the story of the Westcar Papyrus with that of the Memphite Theology and which, in turn, explains the curious variation in the Osirian myth, his 'drowning' in the Nile at the exact spot where the dividing line runs through Ayan and near Abusir.

Skyglobe 3.5 shows that, in epoch 2300BC, which fits the Fifth Dynasty according to the latest chronology, the sun approached the Milky Way from the west, and reached the western shore in early May (Julian). The sun 'drowns' in it for about twenty-four days, to emerge on the eastern shore at the end of May (Julian). At that moment it is in line with the calculated rising time of the 'head' of Osiris, those three little stars which I correlate to the three little pyramids at Abusir on the Memphis-Duat map. The horizon thus joins both the 'head'of Osiris-Sahu and the place where the sun 'emerges' from the waters of the Nile. This astronomical evidence suggests that there was, indeed, an attempt to solarise the cult of Osiris and possibly Osiris himself.[37] Clearly the astronomical evidence brought by the Heliopolitan priests was the concurrence of the 'drowning' of the sun in the celestial Nile and the appearance of the 'head' of Osiris-Sahu. The name of one of the Fifth Dynasty kings who placed his pyramid at Abusir, Sahu-ra, indicates an attempt to merge, and possibly to control, the Osirian astral cult by the Heliopolitan solar faction. It seems to have worked until the end of the Sixth or even the Seventh Dynasty, but the Osirian cult re-emerged, with even more cogency, in the epoch known as the Middle Kingdom which came after the Pyramid Age.[38]

The other version of the death of Osiris was his being killed by Seth and his body cut into pieces and thrown all over Egypt. The six pyramids of the Fifth Dynasty,[39] together with the seven of the great Fourth Dynasty, gives a total of fourteen which comprised the Memphite Necropolis at the time the Pyramid Texts were written. Interestingly, this number ties in closely with another specific aspect of the death of Osiris under the knife of Seth; Seth cut up the body into fourteen pieces.[40] As Wallis-Budge pointed out:

> later tradition asserts that the body of Osiris was cut into fourteen or fifteen pieces, and that over the place where each was buried Isis caused a sanctuary to be built . . . these tomb-chapels, or funerary temples of Osiris may represent the Aats (the Elysian Fields) of Osiris mentioned in the Pyramid Texts . . . the

tombs of Osiris on earth had their counterparts in heaven . . .[41]

Returning to the epic quarrel between Horus and Seth, which followed the death of Osiris, we are told in the Pyramid Texts that in the great fight which took place Horus 'lost his left eye'.[42] This curious mutilation can also be explained by precession. In all sky mythologies and especially in the Egyptian one, there always existed a great bull in the sky represented by the vast constellation of Taurus.[43] This celestial bull is closely connected with Orion the Hunter, such that classical depictions generally show Orion's left arm extending with his hand up to the 'head' of Taurus. Recently, it has been recognised that the Mithraic bull, slain by the Persian-Roman deity, Mithra, is offering an astronomical scene where Mithra is Orion and the 'head' of the celestial bull is none other than the Hyades.[44] This imagery conforms with the classical Greek and Roman representation of Orion and Taurus, with the Hyades being the 'head' of Taurus.[45] It is therefore interesting to note that the 'eyes' of the bull were Aldebaran and star 311 (Epsilon Taurus), the latter being the 'left eye'.[46] We have shown how star 311 crossed the celestial equator going from the lower sky to the upper sky in *c.* 2450BC. Was it then that Horus, who was allocated Lower Egypt, 'lost his left eye'?

The Pyramid Texts make it clear that the epic duel where the 'eye' of Horus was lost is seen as occurring in the lower eastern sky, on the banks of the Winding Waterway:

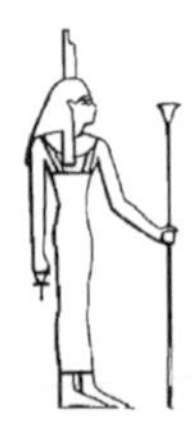

Horus has cried out because of his eye, Seth has cried out because of his testicles, and there leaps up the eye of Horus, who has fallen on yonder (right) side of the Winding Waterway . . . Thoth (the planet Mercury) saw it on yonder side of the Winding Waterway . . . the eye of Horus fell on Thoth's wings on yonder side of the Winding Waterway, on the eastern side of the sky . . . [PT 594–6]

The celestial location is again somewhere near Orion. We read that when Osiris was knocked down near the banks of the

Nile, his assailant, Seth, accuses Osiris in front of the gods of having started the fight: 'It was he [Osiris] who attacked me . . . when there came into being this his name of Orion, long of leg and lengthy of stride . . .' [PT 959].

Thoth and Horus then come to help Osiris to the sky:

> Horus comes, Thoth appears, they raise Osiris from upon his side and make him stand . . . Raise yourself, O Osiris, Isis has your arm, O Osiris; Nephthys has your hand, so go between them. The sky (the Duat sky-region) is given to you, the earth ('Egypt' the Duat of Memphis) is given to you, and the Field of Rushes, the Mounds of Horus, and the Mounds of Seth . . . [PT 956–61]

Are the mounds of Horus and of Seth allusions to the pyramids?

In the British Museum there is a magnificent document dated to the New Kingdom called the Chester Beatty No. 1 Papyrus, where we are given details of what happened in the cosmic courtroom of the gods.[47] It seems that the case had been going on for several years before the 'Heliopolitan Council' and the gods, angered by the long quarrel, were about to give their final verdict.[48] It must have been a difficult one, for in the Chester Beatty No. 1 Papyrus much is made of the efforts surrounding the handling of the matter and how the Egyptians showed 'the triumph of legality over brute force'.[49] It all points to an awkward decision on how the 'Two Lands' previously ruled by Osiris be divided between the two kings after what was probably an indecisive battle. Seth is persuaded to abide by the decision of the Heliopolitan Council. As Jane Sellers previously concluded: 'an important court decision gave the office of Osiris to Horus, and Seth was banished to a position bearing the "southern" constellation ORION' – that is the Hyades stars.[50]

I felt I had reached the end of the stellar investigation. In view of the huge amount of textual and archaeological evidence, there were still many loose ends, but the thickest veil, over the Memphite Necropolis, had been removed and I could now discern a complete stellar plan, executed with poetic elegance and grandeur. I was starting to think in the dual mode the ancients of the Pyramid Age had developed so well: a capacity to think in terms of the sky and the land, using the medium of allegories and symbols to express the combined vision. When the Duat was conjured it came in two blended images, one of the sky and the other of the land. The Nile merged with the Milky Way, and stellar alignments and positions projected themselves on the Memphite Necropolis, over the meridians and latitudes which gridded the clusters of pyramid fields.

However, the articles I had published in 1989–90 made no allusion to the wider vision of the master plan which I had now exposed. It was August 1992, nearly nine years since the whole affair had changed the course of my life. I wanted others to know what I had found, and academic articles do not bring discoveries to a wide audience. Egyptologists had a reading backlog of ten years, in some cases twenty, with hundreds of articles, theses, dissertations and papers all waiting to be reviewed. And even then, nothing much came out of all this material. So I took the big decision: to write a book which would popularise the new ideas and highlight the exciting revelations.

When I told Michele, she heaved a big sigh. For years the family had followed my personal quest; the children had grown in the shadow of 'Ancient Egypt', telling the other kids at school that daddy worked with King Tut when asked what my job was. Luckily I'd had some good engineering consultancies during the years, and the favourable sale of our property in Sydney, which had trebled in value in three years, meant that we could be financially solvent for another eight months, a year perhaps by stretching the bills a little. It was now or never. Michele sighed again and nodded with a smile.

A good 386 computer with a 40 MB memory, a new word-processing program, the conversion of a spare room into an office and the book was on its way. I felt happy and sure that this was the right thing. I put aside the traumas and worries all new authors face, the doubt of ever getting published and the terrible lacunae when words simply do not come, and forged on. By November I had almost completed a first draft. Then, needing a specialised book which I thought might be found in Oxford, I gave myself the day off and drove my little Mini-Rover to that stimulating city.

9 INTERMEZZO AT THE PYRAMIDS

Imhotep, the architect of Zoser . . . is credited by Manetho with having been the inventor of the art of building in hewn stone . . . his achievements became legendary among later generations of Egyptians, who regarded him not only as an architect but as a magician, an astronomer, and the father of medicine . . . [and] . . . the Greeks with their own god of medicine, Asklepios

– I. E. S. Edwards, *The Pyramids of Egypt*

Of these (the Fourth Dynasty kings) the third was Suphis, the builder of the Great Pyramid, which Herodotus says was built by Cheops. Suphis conceived a contempt for the gods, but repenting of this, he composed the Sacred Book, which the Egyptians held in high esteem.

– Manetho, *Aegyptiaca* (epitome), according to Eusebius

I A Meeting of Ways

Driving to Oxford on that cold morning, I had the heater on full blast in the little car. At last I was rid of a persistent lung infection which I had apparently caught while working in Kashmir in northern India. I had gone there on a short consultancy, which was to be my last engineering job before getting down to writing the book. It had taken several months of antibiotic treatment to clean out the infection, but now, on the M40 motorway to Oxford I felt fine.

I wanted to buy a copy of the *Hermetica*, an ancient collection of texts written in Alexandria sometime in the Second Century AD by Greek-Egyptians who ascribed their works to Hermes Trismegistos, supposedly the Ancient Egyptian wisdom god Thoth, inventor of hieroglyphics and science.[1] The last English translation was in 1924 by the Hellenic historian, Walter Scott, and I was hoping to find a copy in a second-hand bookshop in Oxford. As it turned out, I was out of luck; their last copy had gone years ago, and no one else had any either. Just as I was leaving the bookshop, the young assistant called me. Apparently his computer had shown that a small publisher in Dorset, Solos Press, had brought out a paperback edition only a few days previously. I took the address and said I would order the book direct.

Solos Press is owned by Adrian Gilbert. He had started it two years earlier and had published four books, one of which he had written himself.[2] His speciality was the re-issue of unusual books which had been out of circulation but were still in demand by specialised readers, and his last project had been a new edition of Scott's *Hermetica*. I phoned his distributors and asked for Adrian's home number but by coincidence he happened to be at his distributors when I called. As we talked, we found that we had a lot in common. Like me, he had long been interested in Ancient Egyptian religion and especially the pyramids. Inside this new edition of *Hermetica*, Adrian had written his own long Foreword about Ancient Egypt, and I was intrigued by some of his comments and the link he saw between the *Hermetica* and Ancient Egyptian texts. We happily exchanged views on this fascinating subject until I realised that we had been talking for nearly an hour. I told Adrian about my forthcoming book and asked him whether he would be interested in publishing it. He said he would like to see the text and we agreed to meet soon.

We met early in December, and within an hour we had decided to co-author a series of books, the first of which would be *The Orion Mystery*. Adrian's experience was invaluable: he quickly worked out a writing plan and in a week or so *The Orion Mystery* was on its way. I had been working in isolation far too long, and Adrian brought fresh energy and impetus that got the project off the ground. Being now a team, we decided on further lines of research that we could share between us, so that other neglected aspects of the stellar religion could be developed. There were two major lines which needed study and linking to the mainstream of the thesis: the mysterious relic, the Benben Stone in the Temple of the Phoenix, and a development of the precessional star-clock effects of the shafts in Cheops's pyramid.[3]

Soon afterwards, Dr Edwards telephoned to ask if I had done any studies on the shafts in the Queen's Chamber of Cheops's Pyramid. I told him that I had published an article in *DE* 16 two years before. He was eager to know what I thought about them and I told him that my precession calculations showed that the southern shaft had been targeted towards Sirius and that my conclusions, therefore, were that neither the shafts nor the Queen's Chamber was abandoned. He said that he wasn't too sure about that, and was going to send me an article he had written on the subject which was due for publication the following year.[4] Then he suddenly told me that a German scientific team, working under Dr Rainer Stadelmann of the German Archaeological Institute in Cairo, were exploring these shafts at present. This was important news and I decided to take a trip to Egypt as soon as possible. Adrian and his wife, Dee, an amateur photographer who was already working on the pictures for our book, decided they would come too. We planned to leave around the end of February 1993, which would give me time to arrange for the interviews I wanted to secure in Egypt. I was especially keen to meet Dr Stadelmann and find out more about the shaft; I was, of course, hoping to get new data on the slopes and solve the discrepancies which Petrie's reading were giving.

By the end of February Adrian and I had completed the first draft of *The Orion Mystery*. We were not to know that unexpected events would soon force us to rewrite it. We needed a break and it seemed this was a good time to go to Egypt and do some fieldwork as well as taking photographs for the book. A formal meeting with Dr Stadelmann had been arranged by my contacts in Cairo for the first week of March. The international press had been rumbling on about fundamentalist terrorism in Egypt, but the problem seemed to be mostly in Upper Egypt and Cairo was quiet. I telephoned my cousin in Egypt, Josette Orphanidis, and asked if tourists were restricted in that area. Quite the contrary; the Egyptian authorities were anxious to play down the problem and were bending over backwards to make sure that tourists enjoyed their visits. And because the volume of visitors had dropped radically, the archaeological sites were free of the usual crowds. It was the ideal time to go to the Cairo area, and especially propitious for the itinerary we had in mind. Off we went from London on 26 February.

II A Fateful Meeting at the Isis-Sirius Shaft

Adrian and Dee stayed at the old Victoria Hotel, a *grande époque* hotel in the busy district near Ramses Square; I stayed with Josette and her husband, John, in the quieter residential district of Maadi. The weather was glorious and not too hot. We were in high spirits and Dee was eager to get her first shots of the Giza pyramids.

But the first shot was not from Dee's camera but a real explosion at Tahrir Square, not far from the Egyptian Museum. Terrorists had placed a powerful bomb in a small and crowded coffee bar where tourists and, more especially, students from the American University in Cairo gathered for their midday break. Two tourists were killed and fourteen others, mostly local people, were badly injured. We were advised not to wander about in central Cairo, and to stick to the major tourist sites which would now certainly receive the best security Egypt could muster. It would be even better to keep away from the crowds and do our own thing, which is what we had in mind anyway.

The first thing we did, of course, was to visit Giza. It was one of those glorious Cairene days when the weather is tender: a gentle, warm sun with limpid blue sky and a soft breeze. We walked for several hours around the pyramid plateau, breathing the soporific air of the desert, rich in oxygen. We started our tour from the south-west, where a high knoll affords the visitor a marvellous view over the whole necropolis. To the north-east were the three giants, with Menkaura's closest and the Great Pyramid farthest away. Even from this distance, a kilometre or so, it was awesome. We spoke little, preferring to 'listen' to the monuments.

We walked down to the third pyramid, the smallest. Standing before its south face, we were confronted by a wall of stone, and our heads had to be tilted back to see the sky above. The two larger pyramids disappeared from sight and, without their size for comparison, the third pyramid was a giant in its own right. We climbed one of the little satellite pyramids behind us, and sat there taking in the full effect of the massive enterprise. We then walked to the east face, and the third pyramid was dwarfed as the two others loomed before us. We passed the east temples with their huge blocks of stone, some apparently weighing over 200 tons, and admired the fine jointing. Then, after a ten-minute walk we were at the foot of the second pyramid, Khafra's masterpiece. It is impossible to explain the effect this has on the human mind; so many times I had stood there and each time I was awed, humbled and then exhilarated by the sight that towered towards the heavens. We decided not to enter it but to proceed to the main objective: the Great Pyramid and its mysterious shafts.

We clambered the few courses on the north face of Khufu's (Cheops's) pyramid which led to the Ma'amoun entrance, and in we went. We went to the foot of the ascending passage and looked up: before us was a long and dimly lit rectangular tunnel shooting high into the pyramid. Crouching in awed silence, we began to climb deep into the monument. After what seemed like an endless journey, we reached the junction of the Grand Gallery and the horizontal passage which leads to the Queen's Chamber. I looked at Adrian and, almost out of breath, said 'Isis'. He nodded. Crouching again, this time on a level floor, we made our way towards the Queen's Chamber. It was 27 February 1993; in eight days from now Rudolf Gantenbrink would be making the same crouched walk carrying the metal case containing the tiny robot, and starting his exploration of this shaft.

The Chamber was empty of tourists – a rare occurrence, but with the bomb at Tahrir Square, they were staying in their hotels. We stood there, the three of us, and gazed at the walls, at the vaulted ceiling and the great 'niche' on the east side; then I pointed to the opening of the southern shaft. Only three weeks later Gantenbrink would make his historic discovery.

III A Robot and a Door

During the next few days we visited a kaleidoscope of ancient sites – Saqqara, Dashour, Abusir, and also the bustling bazaars of old Cairo. At Saqqara an old *reis* and friend of mine, Ibrahim, complained of the drop in tourists, the gauge of his weekly income. *Maalesh*, I told him: that magical, all-soothing word which loosely translates as 'never mind, it doesn't matter'. We made his day by taking a special tour to the few closed mastabas of the Fifth and Sixth Dynasties in the south-east side of Zoser's pyramid. The walls of these mastabas were covered with exquisite carvings of daily scenes, and with few tourists coming to see them, the ancient paintwork had survived almost unscathed, rich in colour and detail. Here a cow giving birth to a calf, aided by two naked Egyptians; there a mother dressing her young, around her baskets of dates, oranges, melons and figs; young men fishing with spears on reed boats, their catch of Nile perch bursting out of the reed baskets. It was moving to see such vivid scenes of people who lived here more than

4000 years ago, but the atmosphere dissipated when a group of tourists, who had decided to brave the bomb scares, arrived with cameras clicking and guides shouting commentaries and instructions. It was time to go.

I saw Dr Stadelmann on 2 March. A charming and friendly man in his forties, he was frank about the work inside the Great Pyramid. He explained that it had been started early in 1991, under the project team leader, Rudolf Gantenbrink, an engineer and specialist in robotics, and that it consisted mainly in improving the ventilation of the Great Pyramid.

As we have said, the Cheops Pyramid is unique; not only is it the largest and most geometrically perfect of them all but, unlike the others, it contains an elaborate system of above-ground chambers. It is also the one most visited by tourists and this has had unfortunate consequences which could not have been foreseen when it was built. Each person leaves behind about twenty grams of water vapour in breath and perspiration, and the air inside the pyramid had become unhealthy and overly humid. Not only was this uncomfortable for the tourists, who were paying to go inside, but it was causing leaching of salt crystals inside the passages and chambers, making the porous limestone act like a sponge. In places water was dripping from the ceiling. Salt and minerals within the stone were dissolved by the excessive condensation and seeped through to the surface, forming unsightly growths which would eventually cause flaking. Something had to be done to stop this process before the limestone blocks began to crumble and the whole edifice became unsafe. The task of finding a solution to the problem was put in the hands of the German Archaeological Institute and they called Rudolf in as a consultant to carry out the work.

The most obvious solution to the immediate problem of humidity was to increase the air flow throughout the pyramid. This was not so difficult because there were already in existence two small shafts running from the King's Chamber (the uppermost of the three) through the core of the pyramid to the outside. It seemed likely that once these air-shafts had been cleaned and made to work more effectively, the atmosphere inside the pyramid would improve. Accordingly, Rudolf and his team designed and built a machine called UPUAUT (meaning 'opener of the ways' in Ancient Egyptian and originally the name of a jackal god associated with the dead). This device had a camera mounted on it and could be hauled up and down the shafts by pulleys and cables fixed in the King's Chamber, allowing inspection of the ducts from the inside. Once the debris of centuries had been removed from the shafts, a series of heavy duty electric fans were fitted into them so that fresh air would constantly be drawn up into the pyramid. By this simple means the humidity inside was brought down to the ambient level of the desert outside, making the atmosphere healthier for visitors and safeguarding the pyramid from further deterioration.

This first phase of the work was now complete and Gantenbrink had returned to his home in Munich, to bring back a new robot, UPUAUT 2, to explore the shafts in the Queen's Chamber. Unlike the first robot, it had its own traction system so that it could climb up and down the shafts unaided. It also carried headlights, a laser guidance system and a small video camera, to send back pictures to a monitoring console. UPUAUT 2 was a highly sophisticated robot and looked like a remote-controlled moon buggy. Gantenbrink was not due back until 6 March, and Stadelmann said he would arrange a meeting for 7 March. I said this was cutting it fine for me, since we were leaving Egypt that day. Stadelmann hoped Gantenbrink might see me on the evening of the sixth, but he could not promise this.

IV A Meeting With Gantenbrink

On 5 March, after a long evening walk on the Giza plateau, with the constellation of Orion putting on a wonderful display at the meridian, I left a note at his hotel for Rudolf Gantenbrink. I was hoping he would be able to see me the following evening and give a brief interview.

The next evening my cousin John returned from work in his gleaming white Mercedes, which unfortunately attracted a horde of street beggars. It was not a normal sight in Maadi, as John and I knew; it was a sign of the times, and things were far worse than the Egyptian authorities wanted to admit, even to themselves. John then drove me to Gantenbrink's hotel, where the receptionist said he had just arrived with two colleagues and had left instructions for me to call his room.

Rudolf Gantenbrink is a young and handsome man in his late thirties. He greeted me in friendly fashion and asked me to join him and his team for dinner. With him was a film producer from Los Angeles, Jochen Breitenstein.[5] Gantenbrink explained that they would resume exploration in the southern shaft of the Queen's Chamber the following day; tonight they wanted to relax over a good meal and a few beers. We took to each other immediately, and the conversation, of course, was of Egypt and the Pyramids. We talked of the worrying political situation brewing in Egypt and the pitiful condition of the archaeological sites. Jochen Breitenstein felt very strongly about this, and was depressed that the monuments were suffering from lack of attention and vandalism by tourists who were not properly supervised. Gantenbrink was particularly concerned with the Seti I cenotaph at Abydos and the tomb at Luxor. He said that the wonderful paintings and reliefs, many depicting astronomical scenes, had suffered badly from vandalism and perhaps even more from the alarming increase in humidity. The Seti I tomb, like many others such as the tomb of Tutankhamen, was now closed but little was being done about repairing it because few know how to proceed. Apparently the roof of the cenotaph was slowly collapsing.

Rudolf's interest in Egyptology had begun when he heard of the shafts and realised that his robots could help in this sort of exploration. He had already taken his new robot, UPUAUT 2, about twenty metres up the southern shaft of the Queen's Chamber, showing that the shaft was not abandoned by its original builders. They had stopped the exploration to make some modifications to the machine so as to go deeper into the shaft. Gantenbrink had no idea how deep, but said we might be surprised by what might be found in the end. He wondered what my prognosis was now that he had been told of my astronomical findings. I said that whatever was found would be something to do with Isis and Osiris, something connected with their stellar identity. He smiled and assured me I would be one of the first to know the results. He also promised to send me his new measurements on the slopes of the shafts, and hinted that Petrie's measurements were not quite right. This was exciting news. The data would be ready within a week or so, and he promised to fax me the new reading as soon as he had clearance to do so. We parted with the usual exchange of addresses and hoped to meet again.

It was quite late when we got back to Cairo that evening, though the streets were full of people. This was *Ramadan*, the month of fasting, and the Cairenes loved to come out late at night to 'smell the breeze' of the Nile. I picked up Adrian and Dee and drove through Heliopolis and to the airport. They had returned the night before from Luxor and were eager to tell me about the wonderful sights they had seen. I told them of the meeting with Gantenbrink, and we all agreed that we had accomplished more than we had hoped.

V UPUAUT at the End of the Shaft

I decided to tackle the precession problems with the shafts as soon as Rudolf sent the new measurements. By the end of March, I had sent a few faxes to Gantenbrink reminding him of the data I needed, but had received no replies. I assumed he was busy and would attend to it when he could. Deep in our own research, we forgot about his exploration until we heard on the news on 30 March that a bomb had exploded in the pyramid of Khafra. The story was confusing and it was not clear what had actually happened. I sent a fax to Dr Stadelmann asking if Rudolf was all right, but got no answer. On the first of April I decided to telephone: Stadelmann was not in Cairo and Rudolf was back in Munich. Stadelmann's secretary assured me that it had not been a bomb but a faulty electrical connection which had caused the explosion in the second pyramid. It was then that I received a fax from Rudolf apologising for the delay and giving the measurements for the slopes of the King's Chamber shafts. As I suspected, they were slightly different from Petrie's and consequently from those used by Badawy and Trimble in their calculations.[6] The table shows the comparison.

Shaft	Gantenbrink	Petrie
King's Chamber southern shaft:	45° 00′ 00″	44° 30′ 00″
King's Chamber northern shaft:	32° 28′ 00″	31° 00′ 00″
Queen's Chamber southern shaft:	39° 30′ 00″	38° 28′ 00″

I realised immediately that because all slopes were slightly steeper than previously assumed, the age of the Great Pyramid would prove slightly younger, and I quickly did the calculations. The south and north shafts of the King's Chamber were targeted to Al Nitak (Zeta Orionis) and Alpha Draconis respectively; the south shaft of the Queen's Chamber to Sirius. The dates I got were:

Shaft	Gantenbrink	Epoch	Petrie	Epoch
KC south	45° 00′ 00″	*c.* 2475BC	44°30′ 00″	*c.* 2600BC
KC north	32° 28′ 00″	*c.* 2425BC	31 00′ 00″	*c.* 2600BC
QC south	39° 30′ 00″	*c.* 2400BC	38 20′ 00″	*c.* 2750BC

The conclusion was inevitable. The Great Pyramid was built somewhere between 2475BC and 2400BC, thus an average epoch of *c.* 2450BC. This was news. I quickly called Dr Nibbi and she agreed to take two articles, one in *Discussions in Egyptology*

26 and the other in the following issue.[7]

The real excitement was that Rudolf's latest measurements confirmed that the two southern shafts were built at about the same time and that the top shaft pointed to Al Nitak, the lowest star in Orion's Belt (and not Al Nilam, the middle star), which corresponded to the Great Pyramid in the Orion Correlation Theory. The three shafts now locked in perfectly to the stars and the epoch of *c.* 2450BC. Rudolf had no data yet on the northern one in the Queen's Chamber, but he thought it might be closer to 39 degrees. A quick check suggested the same date of *c.* 2450BC for the centre of the four stars forming the 'head' of Ursa Minor, the Little Bear constellation.[8]

Rudolf had told me of his discovery on the telephone and on 4 April a video tape arrived from Munich. I quickly put the tape in and watched as the robot appeared outside the Great Pyramid. Rudolf put the robot into the opening of the southern shaft in the Queen's Chamber and then guided it with the controls on a worktop inside the chamber. The robot began filming inside the shaft. Slowly and laboriously it climbed, going upwards for about sixty-five metres before coming to a stop. In front of it, clearly visible, was what looked like a miniature portcullis slab, of the sort used by the Egyptians to seal off a burial chamber. Attached to the slab, or sliding door, were two copper fittings, one of which was broken, a fragment of it lying on the floor of the shaft. This last part of the shaft was lined with polished Tura limestone, which as far as we know was used inside the pyramids only for lining chambers and was considered sacred by the pyramid builders. It could also be seen from the movement of the robot's laser beam that the slab at the end of the shaft was not fully in contact with the floor but left a gap of about half a centimetre; there was a triangular chip removed from one corner, providing a tantalising glimpse of a grooved channel and a dark recess beyond. Though not conclusive, the video evidence was that what we were looking at was a hatchway leading, perhaps, to some hidden chamber.

I grabbed the phone and called Rudolf. I congratulated him on the amazing discovery and we discussed the details that I had observed on the video. He was, of course, reluctant to speculate what might be beyond the 'door', but he had trouble hiding his excitement. I told him this was big news and that he should go to the press; in fact, I was really surprised that nothing had come out in the Egyptian papers. He had heard that they were preparing a statement, but he was not sure. I urged him to consider letting me go to the British media and at least make sure that the story named him as the discoverer. I decided that I should try *The Times* or the *Daily Telegraph*. I called the *Telegraph* and got one of the editors, Christine McGourty; an interview was arranged and the story was due to appear on 7 April.

Meanwhile I asked Rudolf if I could show the tape to Dr Edwards, and he agreed. He allowed me to show it to anyone who was interested, provided it was not broadcast on TV channels or photographs taken from it. I contacted Adrian and suggested he come over right away.

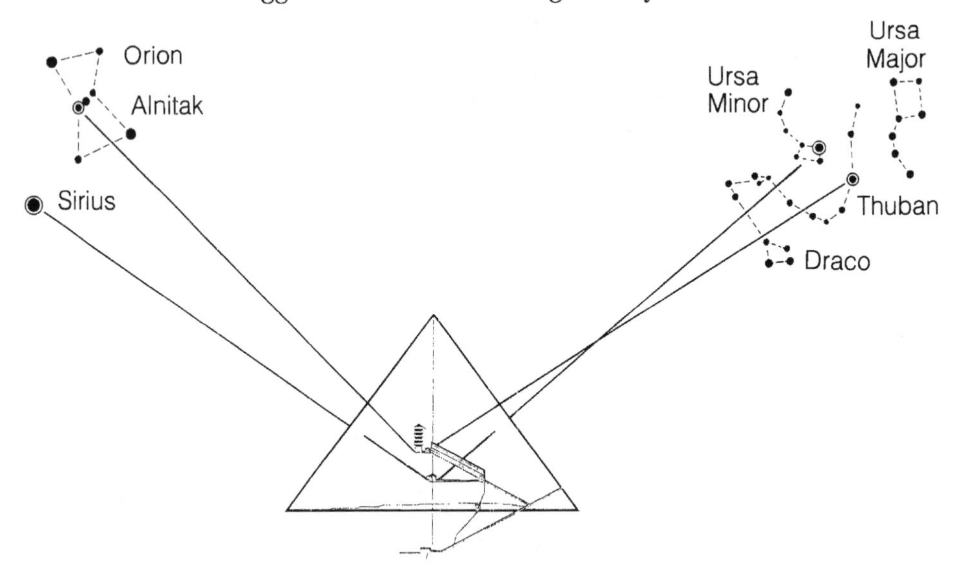

15. *Orientation of the four shafts in the* Great Pyramid

On 6 April, the day before the article came out in the *Telegraph*, Adrian and I arranged to show the tape to Dr Malek and his colleagues at the Griffith Institute at the Ashmolean Museum. They were stunned, and a vigorous debate ensued as to what exactly we had all seen. One thing was certain; there was no doubting the importance of Rudolf's discovery. Even assuming that no chamber existed beyond the slab, this was the first time that any ancient metal has been found inside the pyramid and if the copper fittings on the door turned out to contain more than about 2 per cent of tin, the entire bronze age would have to be redated (Appendix 8). Even the sceptics in our audience couldn't help being excited at the possibility of a find comparable with Tutankhamen's tomb.

We then drove to Dr Edwards's home and showed him the tape. He was thrilled at the discovery, and wondered if more data were available. We called Rudolf in Munich, and he and Dr Edwards had a long conversation. Dr Edwards wanted to see the tape several times, each time picking out new details and asking more questions. He was, of course, keen to know how far the shaft extended above the floor of the King's Chamber. A rough calculation showed that it went about twenty metres above that level, and this alone suggested that the Queen's Chamber had not been abandoned. It was pyramid history in the making, and Dr Edwards suggested that Rudolf should come to England immediately to give a talk at the British Museum.

The next day the *Telegraph* article appeared. It was on page 4 and barely a dozen lines were devoted to the discovery. Dr Edwards said he'd had trouble finding it; surely it deserved more than this? Rudolf was satisfied with the content, but he too was surprised at the small space devoted to it. I contacted Christine McGourty and asked if they were not interested in doing something bigger. She said that the Easter Holidays were coming and most editors were eager to tie up their stories before then, and without pictures there was not much they could do. Rudolf and I agreed that I should go to Munich to discuss the question of pictures that he might provide for me.

Rudolf showed me various video tapes on the shafts and one which he thought would especially please me. It was the filming of the 'Orion' shaft, the southern one of the King's Chamber. It had been taken by UPUAUT 1, and the pictures were breathtaking, the faint speck of light from the outside of the south face of the pyramid becoming larger and larger until it was a sizeable rectangular opening. Rudolf's assistant, who was on the outside of the pyramid, standing precariously on the face of the monument, pulled the robot out and the video camera kept on filming the stunning view of the second and third pyramids below and the Nile Valley to the east. It was for me, in many ways, far more exciting than the film showing the 'door'. UPUAUT 1, in a way the ancient architects would never have imagined, had made the voyage of the soul of Khufu through the narrow shaft leading to the stars.

I returned to England on the tenth of April with six photographs for the newspapers. The *Telegraph* said that they might consider doing something after Easter, but that I was free to try other papers. Eventually the story came out in the *Independent* on 16 April. That same day Channel 4 News contacted me and we arranged to show some of the photographs on the seven o'clock bulletin that evening. Rudolf was interviewed by telephone and Dr Edwards made a live appearance. To our surprise and excitement, when asked what might be behind the 'door', Edwards said he thought there might be a statue of the king gazing out at the constellation of Orion.[9] The Orion Mystery had made it on to nationwide news.

The next few weeks saw us busy with the British Museum conference, which took place on 22 April, one month after the historic discovery.[10] Adrian and I organised Rudolf's arrival with UPUAUT 2 and the technical preparations for the video and slide showing. Many of England's top Egyptologists were there and eager to see the films: George Hart, an expert on Ancient Egyptian religion; Richard Parkinson, a specialist in Egyptian texts; Carol Andrews, a senior member of the profession and old friend of Dr Edwards; T. G. H. James, the previous Keeper; Dr Vivien Davies, the present Keeper, and Dr Robert Anderson, the Director of the British Museum. Such an eminent gathering was a great honour for Rudolf, who returned the gesture by a surprise donation of UPUAUT 2 to the British Museum on the condition 'that I may be allowed to borrow it when the exploration resumes'. Assuring him that the famous robot was in good hands, Dr Davies and his colleagues wished him a successful continuation of his work. There was no speculation about what might be behind the 'door', but there was no doubt that each person there had his or her own idea.

In the meantime, Rudolf had other plans: his ambition was to create a foundation for the preservation and restoration of monuments in Egypt, and he hoped that his high-tech approach was going to rouse the interest and raise adequate funds for this cause. His immediate concern was the alarming deterioration of the Seti I tomb and cenotaph, which was to be the first task of his new foundation. He began the legal and administrative paperwork to create the Upuaut Foundation, and announced that its purpose was to 'do for archaeology what Jacques Cousteau had done for oceanography': popularise a fascinating world which had been forgotten.

Adrian and I were working on a big conference planned for 21 June at the Fédération Nationale des Travaux Publics in Paris. Through this I met Professor Jean Kerisel, who presided at the conference. Kerisel, an active man in his early eighties, is regarded as the grand engineer of Egyptian archaeology. Holder of the Légion d'Honneur, the Croix de Guerre and a long list of titles and important posts in scientific engineering, he is at present the Secretary-general of the Franco-Egyptian Society in Paris.[11] His advice in the coming month would be invaluable.

The Paris conference was a resounding success for Rudolf, with many of the big names in French Egyptology present: Jean-Philippe Lauer, author and expert on the Saqqara step-pyramids; Jean Vercoutter, author and president of the French Egyptological Society, previously the famous Mission Française d'Egyptologie, begun under Napoleon; Jean Leclant, discoverer of pyramid texts in Saqqara and Secrétaire-Perpetuel de l'Academie des Inscriptions et Belles Lettres, where 150 years ago Champollion had made his celebrated announcement of the deciphering of hieroglyphics, and many other prominent members and scholars of the scientific sectors of France. Kerisel had been interested in my Orion Correlation Theory and the recent precession calculations relating to the shafts, and had asked me to show a few overhead slides to Jean Leclant and others who had expressed interest. Leclant seemed to agree that the Pyramid Texts were expressing textually what the Fourth Dynasty had expressed in the astronomical-architectural media of the monuments themselves. Many of the French researchers present felt that the star religion of the Pyramid Texts needed a fresh approach. At last my mission was reaching its goal.

The latest measurements for the shafts confirmed the uncanny accuracy of the builders of the Great Pyramid when they focused on Sirius and Orion's Belt. Since the chances were that they had been aware of precessional changes, they probably knew that the shafts were marking an epoch (*c.* 2450BC). In Egyptian religious texts we often hear of the First Time, when Osiris ruled Egypt during a first golden age.[12] When was this First Time? Could the shafts be used with precession to work this out? And was it to do with the precessional cycle of Orion's Belt?

The excitement surrounding Rudolf's discovery had now to be put behind us as we got back to our own project. We decided to look more deeply into the question of precessional cycles and went back to Skyglobe 3.5 to work out when Orion's Belt began its last cycle.

THE GREAT STAR-CLOCK OF THE EPOCHS

10

We know on the authority of Moses that longer ago than 6000 years the world did not exist . . .
– Martin Luther

The world was created on 22 October, 4004BC at six o'clock in the evening
– James Ussher, *Annals of the World*, 1650

. . . man was created on 23 October 4004BC at nine o'clock in the morning . . .
– Dr John Lightfoot, 1859, the year Charles Darwin presented his work

I The 'First Time' in Ancient Egypt

To know the truth about Egypt's past, we should perhaps heed the words of the wise vizier, Ptahhotep, who lived in the Fifth Dynasty during the Pyramid Age:

> Great is the Truth, enduring is its effectiveness, for it has not been disturbed since the Time of Osiris . . .[1]

Every civilisation has looked far into its mythical past and provided itself with a divine pedigree. For the Greeks this was the Olympian epoch, when the gods fraternised with mortals, as Homer described in the *Iliad* and the *Odyssey*. For the Hebrews it was the time of Genesis and the Patriarchs, expounded in the Old Testament. For the Egyptians, whose civilisation preceded the Greeks and the Hebrews, the first golden age, when gods fraternised with humans, was called *Tep Zepi*, which translates loosely as the First Time.[2]

They believed that the system of cosmic order and its transference to the land of Egypt had been established a long time before by the gods. Egypt had been ruled by a race of gods for many millennia before it was entrusted to the mortal yet divine line of pharaohs. The pharaohs were the sacerdotal connection with the gods and, by extension, represented the link with the First Time; they were the custodians of its established laws and wisdom. Everything they did, every action, every move, every decree had to be justified in terms of the First Time, which served as a sort of covenant of kingship, to abide by and to explain their actions and deeds. This was true not only for the king and his court but applied to all natural events: the movement of the celestial bodies, the unexplained phenomena of nature and the ebbing and rising waters of the Nile. It would not be an exaggeration to say that everything a pharaoh did was connected with the First Time; hence, the careful re-enactment of mythical events which could be either cosmic or secular or both combined in a duality by the power of symbols and rituals. It is not surprising that this blissful First Time was invariably referred to as the Time of Osiris.[3]

The rule of Osiris on earth was seen as Egypt's happiest and most noble epoch and was believed to have been in the distant abyss of time, long before that which Egyptologists are willing to accept as realistic. When the Egyptians built the pyramids, they were thinking of an important event that related to the First Time; whatever that might have been, we now know it had something to do with the stars and, more particularly, the stars of Orion and the star Sirius – the cosmic lands of the souls.

What makes the First Time so interesting is not just that the Egyptians were adamant about its real existence but would pride themselves on being able to compute its epoch, and indeed any epoch in their past. To do that they would need to be aware of precession.

II The Priest-Astronomers of Heliopolis

There had been a tendency to think of the Pyramid Age, and thus the great pyramids, as being of one epoch, one specific dynasty, with a specific group of kings. Yet the enterprise attests something far more grandiose and developed than a temporary surge of creative power during the Fourth Dynasty. All evidence suggests a great plan to freeze time in stone or, better still, to make the stones themselves 'tell the First Time'.

An analogy may clarify the point. A religious monument is often not the expression of its epoch but that the epoch was technically and artistically capable of expressing the origins of a past golden age. When Sir Christopher Wren built St Paul's Cathedral in London in the late seventeenth century, he used modern technology and art in architectural countenance and symbolism which had Christianity as its source. It would be preposterous to suggest that the religion was created by the epoch when the cathedral was built. The same applies to the Vatican Basilica of St Peter's and other monuments. Christianity had its golden age when Jesus roamed the land, and the cathedral is a later epoch's expression of it in the new-found material ability to build such edifices. The religious expression of Wren's or Michelangelo's prowess draws on ideas formulated in the first to the fifth century AD. How old, then, were the religious ideas expressed in the architecture of the Great Pyramid? Centuries, millennia or more? When was the First Time?

We have seen that Gaston Maspero, who discovered the Pyramid Texts, believed that the religious ideas they expressed were several thousands of years older than the version he found in Unas's pyramid.[4] We have seen too that many philologists agree that much of their content is derived from sources going back to pre-dynastic times. Maspero proposed an antiquity of at least 7000 years,[5] but most Egyptologists today find this difficult to accept, claiming that it does not fit the archaeological evidence. Archaeological evidence, however, has proved notoriously faulty, as in the abandonment theory for the Queen's Chamber.[6]

What did the Egyptians feel about the age of their religion? And what did the Greeks, for instance, believe about Egypt's ancient origins?

It has been common sport to pit the Ancient Egyptians against the philosophical 'genius' of the Greeks. Egyptian sages are said to have been but poor relatives to Solon, Pythagoras, Socrates, Plato and Aristotle. As for the sciences of mathematics and astronomy, experts such as Parker and Neugebauer felt that the mathematics was rudimentary calculations children of ten could tackle, and the astronomy simply quaint observation of the stars to interpet superstitious beliefs and the doings of the gods. Whatever skills the Egyptians might have possessed, say these experts, their astronomy was less developed than that of the Babylonians and the Greeks.[7] Yet such views are at odds with what the Ancient Greeks said of the Egyptian sages they made contact with in the early part of the first millennium BC.

Most Ancient Greek and Roman authors believed emphatically that Pythagoras, Plato and even Homer received their philosophy from the Ancient Egyptians.[8] Diodorus (first century BC) tells us: 'The most educated of Greeks have an ambition to visit Egypt to study the laws and principles of a most remarkable nature. Although this country was closed to strangers, those among the ancients known to have visited Egypt: Orpheus, Homer, Pythagoras and Solon . . .'[9]

The great Strabo (64BC–AD25) had this to say:

> The Egyptian priests are supreme in the science of the sky. Mysterious and reluctant to communicate, they eventually let themselves be persuaded, after

> much soliciting, to impart some of their precepts; although they conceal the greater part. They revealed to the Greeks the sécrets of the full year, whom the latter ignored as with many other things . . .[10]

In his famous *Histories*, Herodotus (*c.* 485–425BC), tells us:

> It is at Heliopolis that the most learned of the Egyptians are to be found . . . all agree in saying that the Egyptians by their study of astronomy discovered the solar year and were the first to divide it into twelve parts, and in my opinion their method of calculation is better than the Greeks . . . The name of nearly all the gods came to Greece from Egypt . . .[11]

Dion Chrystomenos (AD30) also pointed out: 'The Egyptian priests much mocked the Greeks because, on many things, they have never known the truth . . .'

What seems to be clear is that the Egyptian priests were regarded by the Greeks as the keepers of great astronomical wisdom which it was not easy to persuade them to divulge to strangers, whom they regarded as unworthy of their high levels of culture. Indeed, strangers entered Egypt only with great difficulty in ancient times – and presumably even greater in the Pyramid Age. In the days of the Fourth Dynasty the primitive Greeks would have appeared as barbarians and other Europeans as no more than cave men to the sophisticated and technologically advanced Egyptians who built the great pyramids. It was not until the Saite Period (*c.* 663BC) that foreigners were allowed to enter Egypt freely,[12] and learn its mysteries.

Schwaller de Lubicz, the modern philosopher, spent most of his life showing that Ancient Egypt was the true repository of philosophy and astronomy (which he termed 'sacred science'). He was convinced that modern scholars are simply not reading the ancients right and that 'there are many revisions to be brought to our judgements regarding ancient peoples of whom only traces remain'.[13]

However, in a letter I got from a prominent Egyptologist working in Cairo, I was told

> We have not the slightest proof that they [the pyramid builders] had any theoretical or systematic knowledge of mathematics. They [had a] really cute [sic] method of doing arithmetical operations . . . I imagine they took the yearly [Nile] inundation for granted . . . In my opinion it's in vain that we look for any mystery in the pyramids, for any secret message left in their texts . . .[14]

To us it is obvious that there is a great mystery here, and that it is time to brave the barrier of experts and try to discern its meaning and message.

III Who Speaks for Ancient Egypt?

Schwaller de Lubicz pointed out that 'there has never been a greater distance between consciousnesses than there is in our time between Western mentality and the mentality of the Ancient Egyptian sages.'[15] Kurt Mendelssohn, who had studied the Egyptian pyramids for many years, put it this way:

> The main difficulty which Egyptologists face today is . . . the state of mind of human society 5000 years ago . . . although man's spiritual world-picture has changed beyond recognition, the laws of physics remain unaltered . . . the knowledge that these same laws were operative and had to be obeyed 5000 years ago . . . provides a reliable link between the pyramid builders and ourselves.[16]

One of the laws of physics that could be most useful in the understanding of the past is, of course, the Precessional motion of our planet and its effect on the apparent position of the stars.

The view now among Egyptologists – and indeed among all students of history – is that dynastic Egypt began *c.* 3100BC. Before this epoch everything is referred to as pre-dynastic, and, as far as general textbooks are concerned, Egypt may as well have not have existed before then. We are told that the first king of Egypt was Menes who unified Egypt in about 3100BC and set up his capital at Memphis. But the concept of dynasties was unknown to the Ancient Egyptians; as they saw it, there had always been, from the First Time, a line of divine kings, the Horus-kings, rightful heirs to the kingdom established by Osiris. The epoch of the First Time was always perceived as going back well beyond the reign of Menes.

From the beginning of scientific Egyptology, which is considered to have begun with Champollion's deciphering of the hieroglyphs in 1822, there was confusion as to when Menes's reign had begun, let alone the age of religious ideas. Champollion placed the epoch of the First Dynasty at *c.* 5867BC, and we have listed the refinements which brought it to 4400BC. Brugsch's system of chronology, based on three generations per century, was again drastically 'refined' to *c.* 3400BC; the date has finally settled around *c.* 3100BC in most of today's textbooks. The technical reasons for all this hopping about since Champollion's estimates are too tedious to review here. They were a mélange of textual analysis, astronomical calculations, carbon dating and a strong dash of personal guesses. The modern experts would not let the Ancient Egyptians speak for themselves.

The Egyptian source most commonly used was from a native priest named Manetho, probably highly educated, a high priest perhaps, who spoke Greek and lived in Lower Egypt during the reign of Ptolemy II Philadephus (347–285BC). Manetho's work has not survived; we have only the commentaries on it by Sextus Africanus (*c.* AD221) and Eusebius of Caesarea (*c.* AD 264–340). We have therefore to assume that Manetho's royal chronology was derived from reliable native sources. Manetho grouped the pharaohs into thirty houses or dynasties; he also provided the Greek versions of pharaonic names: Khufu became Cheops, Khafra became Chephren, Menkaura

Mycerinos and so on. Until the late nineteenth century, Manetho's so-called King's List[17] was the only dipstick to test Ancient Egyptian chronology. Other sources used later were the Abydos List from the Nineteenth Dynasty, the Saqqara List also from the Nineteenth Dynasty, the Turin Papyrus from the Seventeenth and the mysterious Palermo Stone, which gives the annals of the kings of the first five dynasties.[18] It is Manetho, however, who has most influenced modern chronologists.

Manetho ascribed great antiquity to pharaonic Egypt, and speaks of an epoch long before Menes which is quite mysterious. Sextus Africanus, who commented on Manetho's work, was the first Christian historian who devoted his time to producing a 'universal chronology', most of which is compiled in his Chronographiai, which covers the time of 'creation' to AD221. Africanus naturally relied on the Bible as the foundation of his dating, and attempted to synchronise the chronologies of ancient Egypt, Chaldea, Greek mythologies and Judaic history with the new visions of Christianity. The chronological cocktail he produced, thickly laced with bias, can hardly be imagined. Eusebius of Caesarea was the personal chronicler of Constantine the Great, champion and founder of Roman Christianity, so perhaps a little bias is involved there too. Eusebius was more concerned with the formulation of a theory to make history conform with the Christian views of Constantine and prove the validity of the deification of Constantine as Christianity's first imperial saint. In short, both Sextus Africanus and Eusebius were biased towards biblical and especially Roman Christian views of history.

According to Eusebius, Manetho's chronology showed three distinct epochs before Menes: the rule of demigods followed by the Horus-kings, together lasting 15,150 years; then a pre-dynastic line of kings lasting a further 13,777 years: this meant 28,927 years before Menes. Such great antiquity, and thus wisdom, bothered Eusebius. He therefore concluded: 'The "year" I take however to be a lunar one consisting of thirty days: what we now call a month the Egyptians used to style a year.'[19] In this way, Eusebius compressed 28,927 years into 'lunar years' and reduced those before Menes to 2206. Diodorus of Sicily, on the other hand, gave a total of 33,000 years before Menes.[20] But perhaps more significant are the comments in the Turin Papyrus, an original Egyptian document dating from the Seventeenth Dynasty (*c.* 1400BC). It was found in Egypt in the early nineteenth century, and was sold to the Turin Museum in Italy. The third epoch before Menes cannot be deciphered due to damage where the period is given; the two other epochs are listed as of 13,420 years and 23,200 years, a total of 36,620.[21] Egyptologists dismiss much of this as reference not to historical but to mythical epochs. So, were the Ancient Egyptians and later the Greeks wrong about the antiquity of Egyptian civilisation?

We know that Cro-Magnon man, the earliest example of *homo sapiens* or modern man, came on the arena of species evolution about 50,000 to 100,000 years ago. Scientific evidence suggests that the size and shape of Cro-Magnon man's brain was similar to that of modern man. Yet only 134 years ago Charles Darwin was viciously ridiculed for his 'heretical' theory of evolution, and aroused anger from the experts and clerics who maintained that the world had begun with Genesis, in *c.* 4004BC.

'I laughed . . . till my sides were sore', wrote Adam Sedgwick, a British geologist, in a letter to Darwin intended to ridicule his theories. Samuel Wilberforce, bishop of Oxford, declared before the British Association of Science that Darwin's theory was a 'rotten fabric of guess and speculation', and Louis Agassiz, a renowned Professor of Geology and Zoology at Harvard University, cried, 'I trust I will outlive this mania'.[22] How old then was Creation, according to some contemporaries of Darwin?

Dr John Lightfoot, vice-chancellor of Cambridge University, wrote in 1859 that 'man was created on 23 October 4004BC at nine o'clock in the morning'.[23] A century later, scientists agreed that our planet was at least 4.5 billion years old and that hominids, the ancestors of humans, had lived over one million years ago. Then in 1979 the paleoanthropologist Mary

Leakey found a footprint preserved in volcanic ash, believed to be the footprint of an early hominid, possibly an ancestor of humans, dating to 3.6 million years ago. Yet according to present archaeological evidence, we have moved from cave dwellers to space travellers in little more than 5000 years. Could archaeological evidence again be wrong and could Egyptian civilisation be much older than modern scholars concede?

We have already mentioned the bennu or phoenix bird, and how it provided the Ancient Egyptians with the notion of creation and cosmic cycles related to the stars.[24] It seems it was the phoenix, returning after a long period of absence, who opened a new golden age. R. T. Rundle Clark mentions a period of 1460 years,[25] and in his extensive study of the Egyptian phoenix, mentions this same date and also 12,954 years.[26] Fourteen hundred and sixty years is the Sothic Cycle, which was based on the observation of the heliacal rising of Sirius and its shift of one day every four years in relation to the 365-day calendar, completing a full cycle in $4 \times 365 = 1460$ years.[27] But what are we to make of the vast period of 12,954 years? What cycle was that? Did it also apply to Sirius? For those familiar with precession and its effects, 12,954 years is immediately familiar. It is a half-cycle of precession of about 26,000 years and, so far as visual effect is concerned, denotes the time for a star to reach its maximum and minimal range of altitude/declination change.

Let us take a hypothetical star and assume it started its upward precessional cycle of 13,000 years; imagine that it crossed the south meridian at, say, 12 degrees above the horizon. Every year it seems to have moved a fraction higher, at the rate of roughly 12 arcseconds per year. After a little more than two centuries it crosses the meridian at about 13 degrees altitude and so on. After about 13,000 years it reaches its maximum altitude of, say, 55 degrees above the horizon. It begins to go down at the same rate to reach its minimal altitude of 12 degrees in another 13,000 years, back to where it had started, ready to begin another cycle.

Sellers has demonstrated cogently that the ancients had not

only divided the zodiac into twelve parts but were aware that it took the sun 2160 years to travel through each part or age.[28] The result of 2160 × 12 = 25,920 years, the precessional cycle. This huge period of time, though divided into epochs or ages of 2160 years, and these in turn into 360 degrees or portions of 72 years (72 × 360 = 2160 years), was the fundamental basis of the belief in an Eternal Return of the first golden age. A thorough study led Sellers to make this forceful statement: 'I am convinced that for ancient man, the numbers 72 . . . 2160, 25,920 all signified the concept of the Eternal Return.'[29]

The symbol of Eternal Return was, of course, the phoenix, the fabled bennu, and we have seen how in the Pyramid Age its relic or 'seed', the mysterious Benben Stone, was kept in the Temple of the Phoenix at Heliopolis. More importantly, the stylised replicas of the Benben Stone were placed on top of great pyramids. Could these pyramids – and more especially the great pyramids of Giza – be an omnipotent expression of the Eternal Return, the precessional return? The shafts in the Great Pyramid are a powerful indication that this approach is on the right track.

IV The Eternal Return of the 'First Time'

We tend to think of time as something observed when we look at our wristwatches or clocks or a calendar. Take away these things and how are we to know what time it is? How do we know which year or epoch it is? Unless we are astronomers or keen navigators, most of us don't have a clue.

The ancient astronomer-priests of Heliopolis knew the secrets of time, because they observed and studied the apparent motion of the stars, the moon and the sun. If we did too for long enough, most of us would arrive at a variety of calendrical conclusions: the divisions of hours in a day, the number of days in a year, the number of lunar months in a year. Few would know, however, how to fix a year with a marker so that in, say, four or five centuries someone could use our marker and tell the epoch.

The astronomer-priests of Heliopolis knew how to do this, and this was probably one of the great secrets they kept jealously for themselves and, later, from the Greeks. The secret was the awareness of the precession of stars and the ability to calculate the rate of change for those of Orion, the Hyades and Sirius.[30]

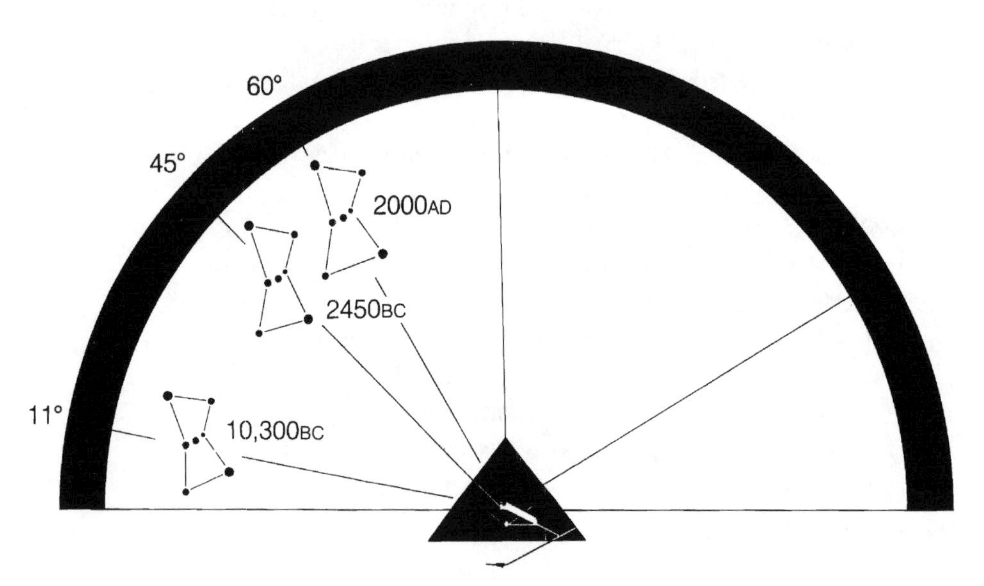

16. *The Position of the constellation of Orion through the ages*

It is customary to attribute the discovery of precessional motion to Hipparchus of Alexandria (*c*. 180–125BC), but many scholars, Zäba, Sellers[31] and Schwaller de Lubicz[32], for example have argued that the Ancient Egyptians had worked it out long before the Greeks and probably prior to the Pyramid Age. We have seen how the Greeks attributed their astronomical knowledge to the Egyptian priests of Heliopolis and Memphis and held that the sages of Heliopolis knew many of the mysteries of the stars. We have also seen how scholars of the Pyramid Texts agree that the stellar cult was an element in the liturgy which might predate the Pyramid Age by several centuries, perhaps several millennia. The Egyptians' special interest was observing the rising of stars and their transit at the meridian, with particular reference to Sirius and the stars of Orion, so it was practically inevitable that they noticed the effects of precession on these stars. As a simple rule of thumb, precession causes a change in declination of just under half a degree per century for these stars. It would have taken a century, two at the most, for the Ancient Egyptians to notice the effects of precession. Taking Zeta Orionis (Al Nitak) to exemplify the case, calculations show that the change in rising point between, say, 3000BC and 2800BC would have been 1.3 degrees of arc as seen from Heliopolis:

3000BC: Azimuth	110.4 degrees
2800BC: Azimuth	109.1 degrees
Variation	1.3 degrees

This is nearly three times the apparent size of the full moon and impossible not to be noticed by stargazers who constantly recorded the rising of stars. If the observations were made at the meridian transit, the apparent variation in altitude over the horizon would have been:

3000BC: Altitude	42.5 degrees
2800BC: Altitude	43.5 degrees
Variation	1.0 degree

This gives one degree of change; again, noticeable to the naked eye. Thus if the Ancient Egyptians were aware of the fact that the stars shifted slowly and that this was easily measurable at meridian transit, the conclusion is inevitable: the architect who designed the southern shaft of the King's Chamber in the Great Pyramid and intentionally directed it to Zeta Orionis, knew that this star would eventually change altitude and also knew that the star was 'fixing' a point (*c.* 2450BC) in the great cycle of time.

It seems reasonable to conclude that the architect also knew the rate of precessional change. The Table shows the changes in declination and altitude at the meridian transit of Al Nitak over 13,000 years.

Year	Declination	Altitude at Meridian
AD2550	−1° 50′	58° 11′
AD2500	−1° 50′	58° 11′
AD2000	−1° 54′	58° 07′
AD1000	−2° 59′	57° 02′
1BC	−5° 13′	54° 48′
1000BC	−8° 28′	51° 33′
2000BC	−12° 38′	47° 23′
2450BC	−15° 01′	45° 00′
10000BC	−48° 39′	11° 22′
10400BC	−48° 53′	11° 08′
10450BC	−48° 53′	11° 08′

[Source: SKYGLOBE 3.5]

Looking from Heliopolis, the lowest point marking the start date of that cycle is 10400BC, when Al Nitak had a declination of −48 degrees 53 minutes and it was 11 degrees 08 minutes over the southern horizon at its meridian transit. The highest point marking the end date of that cycle is about AD2550, when the star would stay for a few decades at a declination around −1 degree 50 minutes and at altitude 58 degrees 11 minutes over the south horizon at its meridian transit.[33] But what now emerges from the visual picture of the southern sky at the epoch *c.* 10400BC is this:

The pattern of Orion's Belt seen on the 'west' of the Milky Way matches, with uncanny precision, the pattern and alignments of the three Giza pyramids!

In *c.* 2450BC, when the Great Pyramid was built, the correlation was experienced when Orion's Belt was seen in the east at the moment of heliacal rising of Sirius, the perfect 'meridian to meridian' patterns, i.e., when the two images superimpose in perfect match; this is when we see the First Time of Orion's Belt in *c.* 10450BC.

It cannot be coincidence that such a perfect arrangement of the terrestrial and celestial central portion of the Osirian Duat, Rostau, occurs at the start of the great precessional cycle at 10450BC. Why such a remote date? Why provide us with a precessional marker defined by the southern shaft, that is, the Belt of Orion shaft of the King's Chamber? Why did the architect who designed this shaft and probably the whole pyramid want to draw attention to this remote First Time date of Osiris in *c.* 10450BC?

V The Timaeus: 10450BC

If a stargazer watched Orion's Belt from the region of Heliopolis *c.* 10450BC, and then recorded or marked the altitude at the meridian or rising point on the horizon, he would unwittingly have fixed the First Time of Osiris. Is there any indication that this could have happened?

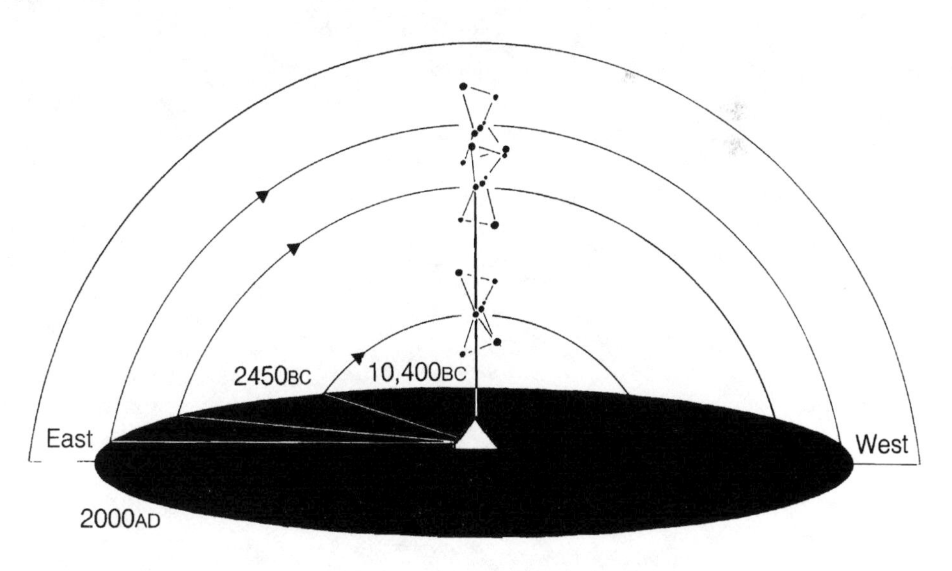

17. *Positions of Rising and Culmination of Orion through the ages*

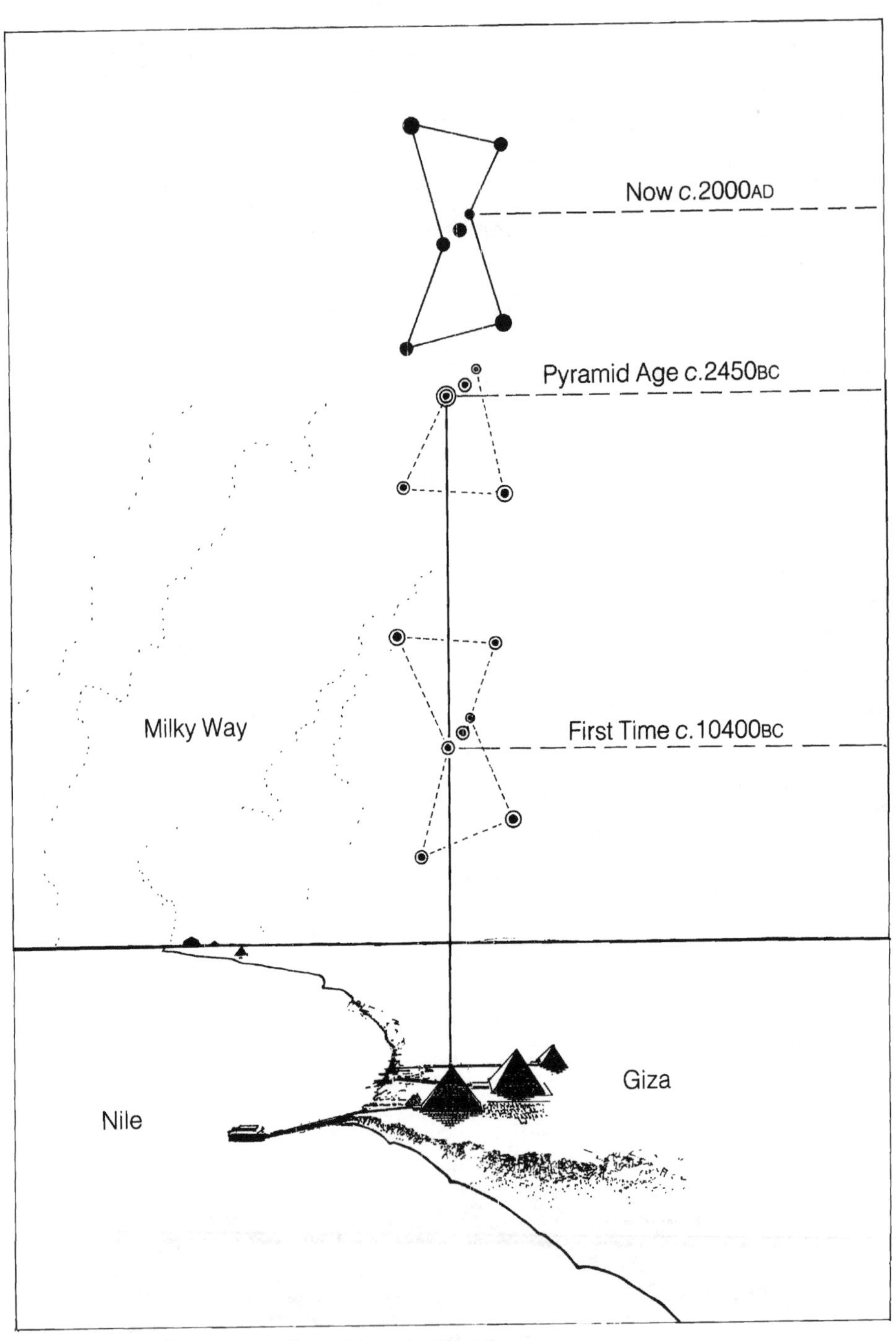

18. ***At around 10400BC the pattern in the sky was mirrored on the ground by the pyramids***

We recall that Strabo wrote in *c.* 20BC, about one hundred years after Hipparchus, that 'the Egyptian priests are supreme in the science of the sky' and that it was they who had 'revealed to the Greeks the secrets of the *full year* [emphasis added], whom the latter ignored as with many other things . . .'[34] Herodotus, writing *c.* 450BC, about three hundred years before Hipparchus, said that it was 'at Heliopolis that the most learned of the Egyptians are to be found . . . all agree in saying that the Egyptians by their study of astronomy discovered the solar year and were the first to divide it into twelve parts . . .[35]

The question has to be asked: was the Giza Necropolis and, specifically, the Great Pyramid and its shafts, a great marker of time, a sort of star-clock to mark the epochs of Osiris and, more especially, his First Time?

We are, of course, aware that 10450BC is far too remote for archaeologists and Egyptologists to entertain, but these findings challenge them to explain – or dispute – the mounting astronomical evidence.

Readers of the Greek classics will undoubtedly bring to mind the Timaeus dialogues of Plato, where he revealed the tragic events of the lost civilisation of Atlantis. The story is reported to Plato by Critias, who said he got it from Solon when he visited the city of Sais in Lower Egypt.[36] It had been told to Solon by Egyptian priests who said that mysterious people from a place called Atlantis had invaded much of the Mediterranean basin as well as Egypt some 'nine thousand years' ago, and that records of them still survived in Egypt. Another aspect of Plato's *Timaeus* which has a connection to our thesis is his statement that the souls of humans are the stars and return to those stars when they die. Plato says that the demiurge made 'souls in equal number with the stars and distributed them, each soul to its several star . . . and he who should live well for his due span of time should journey back to the habitation of his consort star . . .'.[37]

There are, too, the so-called Hermetic Texts, written in Egypt around AD200,[38] which are said by scholars to draw heavily on Plato's *Timaeus*.[39] The unknown authors of the Hermetic Texts claimed, however, that their wisdom came from the ancient books of the Egyptians.[40] In Asclepius III of the Hermetic Texts, Hermes (the Egyptian wisdom god Thoth)[41] asks his pupil: 'Did you not know, O Asclepius, that Egypt is made in the image of heaven . . .?'.[42] This question is intriguing, for Asclepius was associated by the Greeks with Imhotep, the legendary sage and astronomer-architect who designed the first step-pyramid at Saqqara. The Ancient Egyptians said Thoth was responsible for the writing of the sacred books kept at Heliopolis, several of which dealt with the secrets of the motion of the stars.[43]

Two researchers in pyramid studies, W. R. Fix and Mark Lehner, have been bold enough to say that the 'Atlantis' events in Egypt are likely to have happened in 10400BC.[44] This deduction is fascinating, because neither author was using astronomy to deduce this remote date; both were alluding to the so-called 'readings' of Edgar Cayce, an American clairvoyant who died in 1945.[45] Cayce[46] insisted that the Great Pyramid was, at least in its design stage, started around 10400BC, and that the lost records of Atlantis would be rediscovered in a 'hidden chamber' in the last twenty years of this millennium.[47] It would seem that Rudolf Gantenbrink is just in time!

Throughout this book we have tried to stay with the facts. But much as we try to resist such unscientific statements, the Edgar Cayce 'readings', seen from the vantage point of hindsight, are eerie, when it is known that he died in 1945 and, as far as we know, never visited Egypt.

We need now to look into the myth of the phoenix and its egg, the sacred Benben of Heliopolis.

THE SEED OF THE PHOENIX 11

The legend of the phoenix transmitted from century to century and from generation to generation, is lost in the dimness of its origins . . .
– Abbate-Pacha, 'Le Phenix Egyptien'

. . . his relatives ordered that his body should be mummified in the best possible way, so that his soul and his intelligence, when they returned some thousands of years hence to seek his body in the tomb, might find his 'genius' there waiting, and that all three might enter into the body and revivify it, and live with it forever in the kingdom of Osiris . . .
– Wallis-Budge, *The Mummy*

I The Flight of the Fire-Bird

One of the strangest and least understood myths of Ancient Egypt concerns the bennu bird or phoenix. A description of the symbolism it was intended to invoke is given by Rundle Clark:

> One has to imagine a perch extending out of the waters of the Abyss. On it rests a grey heron, the herald of all things to come. It opens its beak and

> breaks the silence of the primeval night with the call of life and destiny, which 'determines what is and what is not to be'. The Phoenix, therefore, embodies the original Logos, the Word or declaration of destiny which mediates between the divine mind and created things . . . In a sense, when the Phoenix gave out the primeval call it initiated all those [calendrical] cycles, so it is the patron of all division of time, and its temple at Heliopolis became the centre of calendrical regulation.[1]

This confirms what we suspected, that the notion of the phoenix is closely related to the Great Pyramid as the epoch and timekeeper of pharaonic kingship, both mythical and historical. The shafts from the King's and Queen's Chambers are calendrical in that they point towards specific stars and fixed their precessional and other cycles; the phoenix, on the other hand, was the herald or bringer of these cycles. There is therefore a link between the phoenix and the pyramid as timekeepers of the stars of Orion and, by extension, the 'soul' of the Osiris-kings. In the *Book of the Dead* (Chapter 17) the question is asked: 'Who is he? . . . I am the great phoenix which is in Heliopolis . . . Who is he? He is Osiris . . .', leaving us with little doubt who was the Egyptian phoenix.

The phoenix also had another important function: it was the bringer of the life-giving essence, the *hikê*, a concept akin to our idea of magic, which the great cosmic bird carried to Egypt from a distant and magical land beyond the earthly world. According to Rundle Clark this 'was "the Isle of Fire" . . . the place of everlasting light beyond the limits of the world, where the gods were born or revived and whence they were sent into the world'. Given that the phoenix is closely linked to the soul of Osiris, and is said to come from the 'place where gods are born or revived', its origins beyond the world are, quite clearly, the Duat.

The story of the phoenix was recorded in more prosaic terms by Herodotus when he visited Egypt:

> There is [a] sacred bird called the phoenix. I have never seen it myself except in pictures, for it is extremely rare, only appearing according to the people of Heliopolis, once in five hundred years, when it is seen after the death of its parent. If the pictures are accurate its size and appearance are as follows: its plumage is partly red and partly gold, while in shape and size it is very much like an eagle. They (the Heliopolitans) tell a story about this bird which I personally find incredible: the phoenix is said to come from Arabia, carrying the parent bird encased in myrrh; it proceeds to the temple of the sun and there buries the body. In order to do this, they say it first forms a ball as big as it can carry, then, hollowing out the ball, it inserts its dead parent, subsequently covering over the aperture with fresh myrrh. The ball is then exactly the same weight as it was at first. The phoenix bears this ball to Egypt, all encased as I have said, and deposits it in the temple of the sun. Such is their myth about the bird.[2]

Although told in the usual storytelling style of the Greek chroniclers, Herodotus actually discussed this matter directly with the priests of Heliopolis and we may suppose that he had no reason to alter the facts. What is more likely, however, is that he unwittingly altered the symbolism of how the Egyptian priests themselves understood the attributes of the phoenix. 'Arabia', for example, stood for the 'east', the land beyond the horizon where the sun and stars rise, that is 'the place where the gods are born'. The phoenix came to Egypt to lay its egg, the term 'ball' in Herodotus's tale implies a fairly large specimen. Herodotus also says that it is made of 'myrrh', a resin commonly used in the process of mummification.[3]

What was it that the Egyptians looked upon as the egg or seed of the phoenix which was linked to the soul of Osiris and, consequently, the stellar rituals of rebirth?

As we have said, the Egyptians called the phoenix the *bennu*. John Baines, Professor of Egyptology at Oxford University, pointed out that the root word *ben*, was generally used by the Ancient Egyptians to denote sexual, procreational or seeding ideas, such as 'the semen', 'to copulate', 'to fertilise' and so on.[4] Interestingly, in Semitic languages the word *ben* also means seed in the sense of son.[5] The direct connection between the bennu/phoenix bird and the Benben Stone kept in the temple of the bennu/phoenix has been made in Chapter One. The fact that the Benben Stone was conical has also been established by many Egyptologists.[6] In a very ancient stela dating from the First Dynasty, the phoenix is seen perched on some object, which Rundle Clark called a 'stone perch'.[7] Later it was commonly depicted perched on a pyramidion or a perch fixed on a pyramidion. Various opinions are expressed by Egyptologists as to what or who the Egyptian phoenix was, but the consensus is that it sometimes represented the soul of Ra, at other times the soul of Osiris, and at yet others the 'Morning Star'.[8] Rundle Clark also rightly said that 'the bird and the stone – if stone it is – are linked together',[9] and that Kurt Sethe, the first acclaimed translator of the Pyramid Texts, identified the Benben Stone with the sacred conical stones of the Greeks and Syrians, the 'Omphalos or Baetylos', a term used by historians to signify sacred stones with cosmic attributes.[10] Indeed, in the earliest known depiction of the Benben Stone on which the phoenix is perched, the Stone is not pyramidal, as was previously thought; its slopes bulge a little, showing that it was conical.[11] It is clear, too, that the Benben Stone was considered a relic of immense value by the pyramid builders, so valuable that it was placed in the holy of holies of Heliopolis, in the focal point of the 'Mansion of the Phoenix' and replicas of it placed on the top of great pyramids.[12]

The conclusion must be that the phoenix was a symbol of divine procreation and rebirth, this magical quality characterised by the seed it deposited in Heliopolis. What, then, was the seed of the phoenix?

II The Seed That Fell From Heaven

We tend to think of meteorites as stones that fall from the sky, though 'falling star' and 'shooting star' are still used as visual metaphors. The fall of meteorites is spectacular. Historical accounts are in close agreement that a fiery mass appears in the sky; shooting down, it sometimes leaves a luminous trail, and its fall is accompanied by what is often described as 'thunder'.[13] Meteorites enter the earth's atmosphere at great velocity but are then slowed down by the friction of the air and great heat is generated around the meteorite. This release of heat, which ignites its surface, causes the fireball appearance, the hot gasses which incandesce around it making the fireball appear quite large. As the meteorite tears its way through the air, it also produces shock waves which resound like cannon fire or thunder, which is probably why in earlier times meteorites were associated with storm gods such as Haddad in Phoenicia and Zeus in Greece.[14]

There are two sorts of meteorites: stone and iron. The iron, for obvious reasons, tend to be black and often larger than the stone variety, since they suffer little or no damage when they hit soft ground. Also, when entering the earth's atmosphere, some iron meteorites retain their direction of flight rather than roll about. These are called oriented, that is, they maintain their orientation as they fall, like an arrow or pointed cannon shell. As these oriented meteorites are heated during their fiery fall, their front part tends to melt and taper down and, when found usually have the characteristic shape of a cone. Two good examples are the large conical meteorites known as 'Morito' and 'Willamette'.[15]

There is evidence of religious cults based on the veneration of sacred meteorites in the ancient world. It is well known that the Greeks regarded Delphi as the 'navel' of the world. However, the omphalos stone which marked the spot was not the original fetish of Delphi. There was originally a rough stone, believed to have been cast down to earth by the titan Kronos.[16] The Delphians believed their stone to be the one cast down by Kronos and called it Zeus Baetylos, a term usually taken to mean meteorite by historians.[17] Extant drawings show the Zeus Baetylos as ovoid in shape, and about the size of a cannonball. In view of its cosmic origins and characteristic shape, the Zeus Baetylos was almost certainly a meteorite.[18] A similar stone was shown to the historian Pausanias (second century AD) at the town of Gythium, which the locals called Zeus-Kappotas (Zeus fallen down). This was probably also a meteorite.[19] Pliny (AD23–79) also reported that a 'stone which fell from the sun' was worshipped at Potideae and that others had fallen at Aigos-Potamus and at Abydos, near the Hellespont.[20]

The cult of meteorites was particularly rife in Phoenicia and Syria. At Emessa (Homs), for example, was the shrine of the god Ela-Gabal or Elagabalus, where a sacred relic was described as 'a black, conical stone'; the chronicler Herodianus tells us that the Emessians 'solemnly assert it to have fallen from the sky . . .' Not far from Emessa, in the temple of Zeus-Hadad, at Heliopolis-Baalbek, were 'black conical stones'.[21] Zeus-Casios, a counterpart of Zeus-Hadad, had his abode on Mount Casios and also had a *baetylos* sacred to him. In ancient Phrygia (central Turkey) the Great Mother of the Gods, Cybele, was represented at the temple of Pessinus by a black stone said to have fallen from the sky.[22] The Cybele cult was particularly widespread and was adopted by the Romans who took it as far as France and England.[23]

There are many other examples of meteorite worship in many places of the world.[24] This is quite understandable because ancient man saw the meteorite as the material representation of the sky gods and, perhaps more specifically, the star gods. We surely do not need any further examples to make the point that the Benben Stone kept inside the Temple of the Phoenix may have been a conical meteorite.

That meteorites played a major role in the formation of religious ideas and in the rebirth cult has been known to Egyptologists since 1933. In-depth studies on the subject were made by G. A. Wainwright, a British Egyptologist and former assistant of Flinders Petrie. These appeared in the *Journal of Egyptian Antiquities* from 1933 to 1950. Wainwright traced the evolution of the Egyptian 'meteoritic cult' and its association with several important gods; in particular, he showed that the 'aniconic' (like a cone) form of the Theban god Amun was a meteorite known as the *Ka-mut-f*,[25] quite typical, Wainwright remarked, of small, pear-shaped iron meteorites.[26]

III The Iron Bones of the Star Gods

Although the pyramids were built before the bronze and iron ages, *meteoritic* iron was known to the Egyptians of the Pyramid Age.[27] The Ancient Egyptian name for iron was *bja* and, according to Wainwright, 'meteorites consist of *bja*'.[28] The word *bja* is mentioned repeatedly in the Pyramid Texts in connection with the 'bones' of the star kings:

> 'I am pure, I take to myself my iron (*bja*) bones, I stretch out my imperishable limbs which are in the womb of Nut . . .' [PT 530]
>
> 'My bones are iron (*bja*) and my limbs are the imperishable stars.' [PT 1454]
>
> 'The king's bones are iron (*bja*) and his limbs are the imperishable stars . . .' [PT 2051]

As these passages show, there was a belief that when the departed kings became stars, their bones became iron, the heavenly material (meteorites) of which the star gods were made. Such cosmic iron objects were the only material evidence of a tangible land in the sky populated by star souls, and it was easy to see why the stars were thought to be made from *bja*. Since the souls of departed kings were the stars, they too had bones made of iron.[29]

This brings us back finally to the Benben Stone of Heliopolis, which I[30] and many Egyptologists have associated with a meteorite. Wallis-Budge was the first to suggest that the Benben Stone was a relic similar to the Black Stone of the Ka'aba. The same idea crept into the mind of Egyptologist J. P. Lauer who wrote that the Benben was probably a Bethyl or a meteorite.[31] It is thus quite likely that a large oriented iron meteorite fell near Memphis at some time in the third millennium BC, perhaps during the Second or Third Dynasty. From depictions of the Benben Stone,[32] it would seem that this meteorite was from six to fifteen tons in mass, and the frightful spectacle of its fiery fall would have been very impressive. The fall would have been presaged by loud detonations caused by the shock waves, and even in daylight a fireball with a long, pluming tail would have been visible from considerable distances. This fire-bird would have evoked the notion of a returning phoenix crashing in from the east (according to the *Encyclopaedia Britannica*, all meteorites follow the path of the sun). Rushing to the spot where it landed, people would have seen that the fire-bird had disappeared,[33] leaving only a black, pyramid-shaped *bja* object or cosmic egg (the oriented iron meteorite). They would then have taken it to the ancient temple of Atum, to be placed on the sacred column venerated there.[34]

IV The Seed That Is Osiris

The Pyramid Texts are full of references to the seed of Ra-Atum. The seed in question is that from which Osiris was created in the womb of the sky goddess, Nut, Mother of the Stars: 'O Ra-Atum, make the womb of Nut pregnant with the "seed" of the spirit (Sahu) that is in her. . . . [PT 990] . . . Pressure is in your womb, O Nut, through the "seed" of the god which is in you . . .'

To which the Osiris-king responds, 'It is I who am the "seed" of the god which is in you [PT 1416–7] . . . the Osiris-king is an imperishable star, son of the sky-goddess [1469] . . . O Ra-Atum, this Osiris-king comes to you, an imperishable spirit . . . your son comes to you . . .' [PT 152].

The two-step process of the stellar transfiguration of an Osiris was briefly discussed earlier,[35] where we saw how the corpse was first made into an Osiris-mummiform, then placed inside the rebirth chamber of the sepulchre, where he was to spiritualise himself into a star soul. We learnt that the word for making a mummy in Ancient Egyptian was, not surprisingly, Sahu, synonymous with the name given to the original Osiris when he became the Lord of the Duat.[36] The dramatic act of giving life to the mummy was not expected to happen by itself but depended on the devotion and action of the dead king's eldest son, the new Horus-king who before his coronation was probably called Horus-the-Elder.[37] The crucial dramatic ceremony this Horus had to carry out was called 'the opening of the mouth', which required that the embalmed body of his father, now in full Osirian regalia, be placed upright in front of a small stand on which was a lotus plant in full bloom. The lotus symbolised the 'four sons of Horus' (the king's grandsons[38]), who in turn symbolised the 'four cardinal points'.[39] Wearing a hawk-mask, the Horus slowly approached the mummy and, assisted by his 'four sons', picked up a small metal cutting instrument, similar to a carpenter's adze, and struck or cut open the mouth of the Osiris-king. The four sons, using their 'fingers' (apparently made of *bja*), performed the same ritual. These rites were extremely ancient and are described in the Pyramid Texts:

> 'O King, I have come in search of you, for I am Horus; I have struck your mouth for you, for I am your beloved son; I have split open your mouth for you . . . with the adze of Upuaut . . . with the adze of iron . . .' [PT 11–13]

> '. . . your children's children together have raised you up, [namely] Hapy, Imsety, Duamutef and Kebhsenuf, [whose] names you have [wholly] made. [Your face is washed], your tears are wiped away, your mouth is split open with their iron fingers . . .' [PT 1983–4]

There are three important aspects of this rather bizarre ceremony which demanded our undivided attention. The first was that the adze instrument and also the fingers of the four sons of Horus are said to be made of *bja* (meteoritic iron).[40] This was picked up by G. A. Wainwright in 1931, and discussed in detail in a landmark article entitled 'Iron In Egypt'.[41] Wainwright rightly argued that it was because of the 'heavenly' qualities of *bja* that the ceremony was believed to evoke the magic for the escape of the soul to the stars.[42] This is now a well-accepted notion in Egyptology, and was recently repeated by Dr Bernd Scheel, an expert in ancient Egyptian metal-working and tools, who wrote

> Iron was [a] metal of mythical character. According to legend, the skeleton [bones] of Seth . . . was of iron. Iron was called the 'metal of heaven' because for a long time the Egyptians knew only meteoric iron, which has a high nickel content. Because of its supposedly divine origin, meteoric iron was used in particular for the production of protective amulets and magic model tools which were needed for the ritual called the 'opening of the mouth', a ceremony which was necessary to prepare the mummy of the deceased for life after death.[43]

What Wainwright and also Mercer, the Canadian Egyptologist who translated the Pyramid Texts in 1952, noticed was that the adze used for opening the mouth was shaped in the form of the constellation of Ursa Major, which the Egyptians called *meshtw*, the Thigh.[44] The German Egyptologist, Bochardt, had, however, argued that it was more probably Ursa Minor.[45] A bovine foreleg has the knee bending forwards and thus fits better the shape of Ursa Minor. In any case, these constellations form a pair in the circumpolar region of the sky and are in the region targeted by the two northern shafts in the Great Pyramid. The important cardinal direction for this curious meteoritic and stellar ceremony with the king's mummy was, then, the circumpolar north, the focal point of which is the celestial pole. During the Pyramid Age this was marked by the star Alpha Draconis, the precise target of the northern shaft of the King's Chamber. The northern shaft of the Queen's Chamber pointed to the 'head' of Ursa Minor, made up of four stars which, in all probability, represented the adze used by Horus in the ceremony of the 'opening of the mouth'.

In the Pyramid Texts this instrument is called 'the Adze of Upuaut' [PT 13]. Upuaut was, as we have said, the Jackal-god

19. The Zodiac of Denderah (Ptolemaic Period)
Note constellations of Sahu-Orion (Osiris figure)
preceded by Taurus; the star Sirius (Isis) over ruminating cow
follows Orion

who 'opened the ways' and he is clearly represented in the famous Zodiac of Dendera, now in the Louvre Museum, as the circumpolar Horus figure holding Upuaut. The northern shafts are not only set meridionally but, unlike their southern counterparts leading to Osiris-Orion and Isis-Sirius, they have a curious architectural anomaly, which has perplexed Egyptologists and recently has been queried by Rudolf Gantenbrink, who explored them in 1992–3.

During the conference on 21 June 1993 at the FNTP in Paris, where Gantenbrink, Edwards and I were among the speakers, Rudolf raised the question of this anomaly.[47] He pointed out that when he guided his robot up the northern shafts he came to the junction where they meet with the Grand Gallery. Because the Grand Gallery is in the direct path of the shafts, both had to be given a pronounced 'kink' westward to bypass the Gallery. Rudolf, who is pragmatic and a thorough rationalist, said he could understand that the architects and builders might have made a mistake in putting the opening of the northern shaft in the Queen's Chamber directly in line with the Grand Gallery, and then had to detour to bypass this huge obstacle. What he could not understand, was why this was repeated for the northern shaft in the King's Chamber. He asked the attending Egyptologists – Edwards, Leclant, Lauer, Vercoutter and Kerisel – what they thought of this. Although all are experts on the pyramids of Egypt, no answer was forthcoming. Rudolf then adduced the logical conclusion: the detour or kink was not a mistake but a deliberate design feature. Moreover, these shafts had been given more gentle kinks as they ran past the Grand Gallery and then reverted to their original course.

What was not realised at the time of the conference was that the shafts, with their kinks, appeared to be shaped in the form of the sacred adze. That they were also directed to the circumpolar constellations, one of which symbolised the stellar adze, making this very unlikely to be a coincidence. It now seemed certain that the ceremony for the opening of the mouth had been performed, perhaps several times, inside the

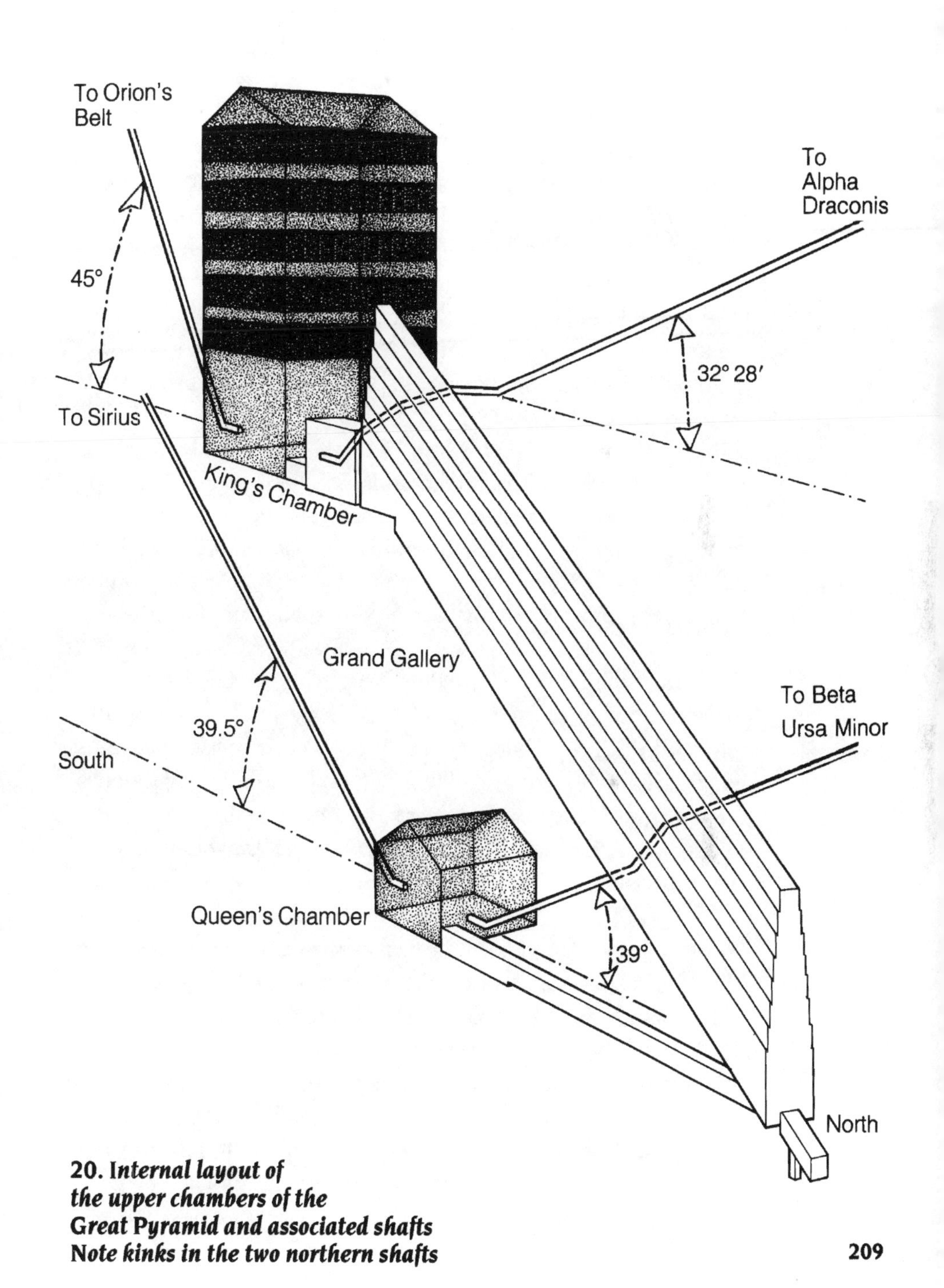

20. Internal layout of the upper chambers of the Great Pyramid and associated shafts Note kinks in the two northern shafts

Queen's Chamber.[48] We could visualise the Horus-son being led into the Queen's Chamber of the Great Pyramid to meet the mummy of his dead father: 'O Horus, this king is Osiris, this pyramid of this king is Osiris, this construction of his is Osiris, Betake yourself to it . . .' [PT 1657]. And Horus exclaiming: 'O King, I have come in search of you, for I am Horus; I have struck your mouth for you, for I am your beloved son; I have split open your mouth for you. I announce him to his mother when she laments him, I announce him to her who was joined to him.' [PT 11–12].

Horus performs the potent ritual then presents his four sons, the dead king's grandchildren: 'I split open your mouth for you . . . I open your mouth for you with the Adze of Upuaut, I split open your mouth for you with the Adze of Iron which splits open the mouths of the gods . . .' [PT 13]. '. . . your children's children together have raised you up, [namely] Hapy, Imsety, Duamutef and Kebhsenuf, [whose] names you have [wholly] made. [Your face is washed], your tears are wiped away, your mouth is split open with their iron fingers . . .' [PT 1983–4].

A priest then acts for the dead king who has been struck by the astral power of the *bja*, and says: 'I am pure, I take to myself my iron (*bja*) bones, I stretch out my imperishable limbs which are in the womb of Nut . . .' [PT 530] . . . 'My bones are iron (*bja*) and my limbs are the imperishable stars.' [PT 1454] . . . 'The king's bones are iron (*bja*) and his limbs are the imperishable stars . . .' [PT 2051].

What now astounded us, however, was to discover whence the Horus was summoned to betake himself to the 'pyramid that is Osiris'. He began his journey towards the pyramid from a place directly due north. According to French Egyptologist Goyon[49] this was precisely on the meridional line of the Great Pyramid, 15.75 kilometres away, the site of the ancient city of Khem, later called Letopolis by the Greeks. It provided the son-priest in charge of the opening of the mouth ceremony with the title, Horus of Letopolis.[50]

Letopolis actually existed before the Pyramid Age,[51] and

many Egyptologists believed it had served as the central geodetic marker for all other sites in the area.[52] This, according to Goyon, was especially the case for the meridional sighting and alignment of the Great Pyramid and, consequently, the whole Giza Necropolis.[53] An even more curious revelation was that, according to Wainwright, the city of Letopolis was the 'Thunderbolt City', so-called because it was linked with a meteoritic cult: '. . . since the Egyptian religion included a very important ceremony of "Opening of the Mouth" of the dead King with tools made from meteorites, it is no accident that the chief "Opener of the Mouth" lived at the thunderbolt city of Letopolis . . .'[54]

In an atlas of Ancient Egypt,[55] we found that Letopolis was, as Goyon had said, about fifteen kilometres due north of the Great Pyramid. What Goyon had *not* said was that Letopolis was also due west of the Temple of the Phoenix at Heliopolis. It thus was on the geodetic point adjoining the meridian of the Great Pyramid and the latitude of the Heliopolis, where the Benben Stone was kept. Letopolis was a signpost to Rostau, the 'roads of Osiris in the sky'.[56] It also linked, by latitude and meridian, the Benben Stone with its stylised replica on top of the Great Pyramid. Finally, it brought together, to meet inside the pyramid, the stellar adze of the circumpolar stars with the roads to Osiris in the sky, which could only be the southern shafts. These lead to the land of Osiris in the sky and, of course, to the Duat.

As if this were not mysterious enough, at the end of one of these southern shafts was the closed door which Gantenbrink hoped to open. Assuming that the door concealed a chamber, might it have been modelled on the secret chamber of Thoth spoken of in the Westcar Papyrus? More particularly, could it contain something of perhaps greater significance than a mummy, a statue or other funerary equipment?

We now turned our attention to Heliopolis, where the Benben 'Phoenix' shrine once stood.

12 THE ROADS OF OSIRIS

I have travelled by the roads of Rostau on water and on land . . . these are the roads of Osiris; they are in the sky . . .
– from the Book of the Two Ways, written on the inside of coffins of the Middle Kingdom, El Bersheh

I Where is the Benben Stone?

Looking at a map of the Memphis-Heliopolis region as it was when the Giza pyramids were built, we see that the position of Heliopolis, and where the great obelisk of Sesostris I (*c.* 1970BC) stands today,[1] is on a line that extends from the south-east corners of the three pyramids of Giza. This was brought to my attention by Dr Gerhard Haeny of the Swiss Institute of Archaeology in Cairo, in a letter he wrote to me in 1986. He said that it had been pointed out to him that the south-east corners of the three pyramids were in alignment and that if that line was extended, it attained the site of the obelisk of Heliopolis. He wondered if this obelisk perhaps replaced an earlier construction.[2]

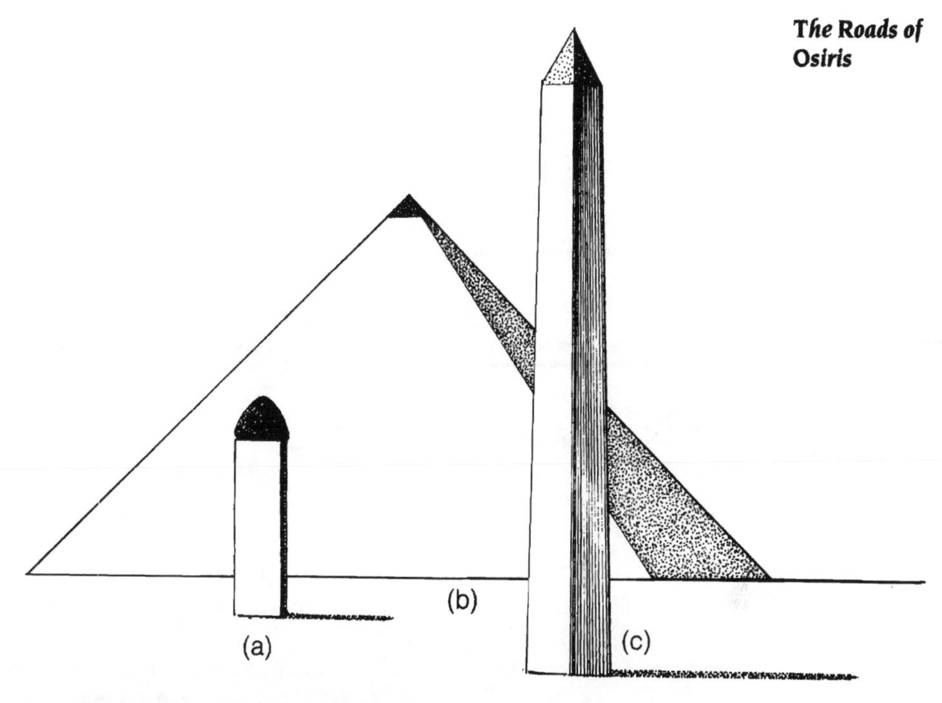

21. *The Benben through the Ages*
(a) The original Benben of Heliopolis as it may have looked
(b) A pyramid surmounted by a pyramidion or Benben
(c) An obelisk tipped by a Benben-T

Actually the Sesostris I obelisk did replace an earlier landmark, and an important and mysterious one at that. Where the obelisk now is at Heliopolis, there once stood the House or Temple of the Phoenix. And in this temple was kept the sacred Benben Stone. Sesostris I, who restored the sacred city of Heliopolis, confirms that his obelisk replaced the Benben Stone – presumably by then 'lost' – for he ordered an inscription to be carved on a stela at Heliopolis: 'My Beauty shall be remembered in His House, My Name is the Benben and my name is the lake . . .'[3]

What Sesostris appears to imply is that the pyramidion or Benben making the apex of his great obelisk was now raised in the house or temple where the original Benben Stone had stood not long before. James Breasted tells us that 'this object was already sacred as far back as the middle of the third millennium BC, and will doubtless have been vastly older'.[4] He adds, 'an obelisk is simply a pyramid upon a lofty base which has indeed become the shaft.'[5] However, many questions remain. Who was Sesostris I? Why was it necessary to mark the place of the Benben Stone with an obelisk? And where had the Benben Stone gone? To answer these questions, we need to look at the history of Ancient Egypt after the Old Kingdom.

It seems that there was much political and social upheaval in the reign of Amenemhet I (*c.* 1990BC), father of Sesostris I. This is attested by several well-preserved papyrus texts, in one of which Amenemhet I gives what at first sight seems rather Machiavellian advice to his son:

> Hearken which I say unto thee, that thou mayest be king of the earth . . . harden thyself against all subordinates, the people give heed to him who terrorises them, approach them not alone, fill not thy heart with a brother, know not a friend, nor make for thyself intimates . . . for a man has no people in the days of evil. I gave to the beggar, I nourished the orphan . . . but he who ate from my hand made insurrections . . .[6]

Yet this terrible pessimism seems to be mitigated by a messianic hope of a return of a 'Great One', expressed by a solitary scribe, Ipuwer, in the reign of Amenemhet I.[7] This text is known to Egyptologists as 'the admonition of an Egyptian sage, Ipuwer', who was undoubtedly a priest at Heliopolis. It is the lament of a sage-priest who finds much confusion at court and in the land. There seems to be total chaos, with the populace entering and defiling the temples once carefully guarded by the priests; holy inscriptions are defaced, departmental offices are raided, and so on.[8] The text clearly refers to the aftermath of a revolution, with the chaos and killings which follow such events: 'Behold, the district councils of the land are expelled . . . a man smites his brother and the same mother. What is to be done?'[9]

The sage-priest is obviously addressing the court, which

seems to be in emergency council and at a loss what to do next.[10] Ipuwer, apparently the only one with the sense and courage to speak, says: 'The districts of Egypt are devastated . . . every man says "we know not what has happened to the land" . . . civil war pays no tax . . . what is treasure without revenue? . . . woe is me for the misery of this time.'[11]

Then he speaks of a great messianic hope, obviously intended for the son of the old and discredited Amenemhet I, who seems to have lost control over the people and the land. Ipuwer calls for a full resumption of the sacred rituals and observances at the temples, and reminds them of the time when an 'ideal king' had ruled Egypt in justice and peace: 'Remember . . . it is said he is the shepherd of all men. There is no evil in his heart . . . Where is he today? Does he sleep perchance? Behold his might is not seen . . .'

Ipuwer makes a strange allusion to something 'concealed' within the pyramid, something he fears might not be there any more: 'that which the pyramid concealed has become empty . . .' Whatever the pyramid concealed was something of great value, indeed something so important that Ipuwer found it necessary to voice a powerful warning about it at court. While we cannot be sure what it was that so concerned Ipuwer, Sesostris I, who seems to have fulfilled Ipuwer's messianic hopes, placed a great obelisk to mark the place where once had stood the most sacred of 'pyramids', the Benben Stone. Perhaps the knowledge of what had once been concealed inside the Great Pyramid had been lost. Certainly, when the pyramid was opened many centuries later by the Caliph Al Ma'a-moun, nothing was found.

However, one further hope remained. Could the genius architect who designed the Great Pyramid have ensured that 'that which was concealed in it', was impossible to find and even more impossible to reach? Impossible, that is, without a little mechanical robot guided by electronic devices?

II Signpost to the Benben Stone

Let us take a look at the geographical environment where this drama may have taken place. The distance from Giza to the supposed position of the Temple of the Phoenix, going north-east, is about twenty-four kilometres. The distance from Giza to Letopolis, going due north, is just under sixteen kilometres, and that from Letopolis to the Temple of the Phoenix, due east, about eighteen.

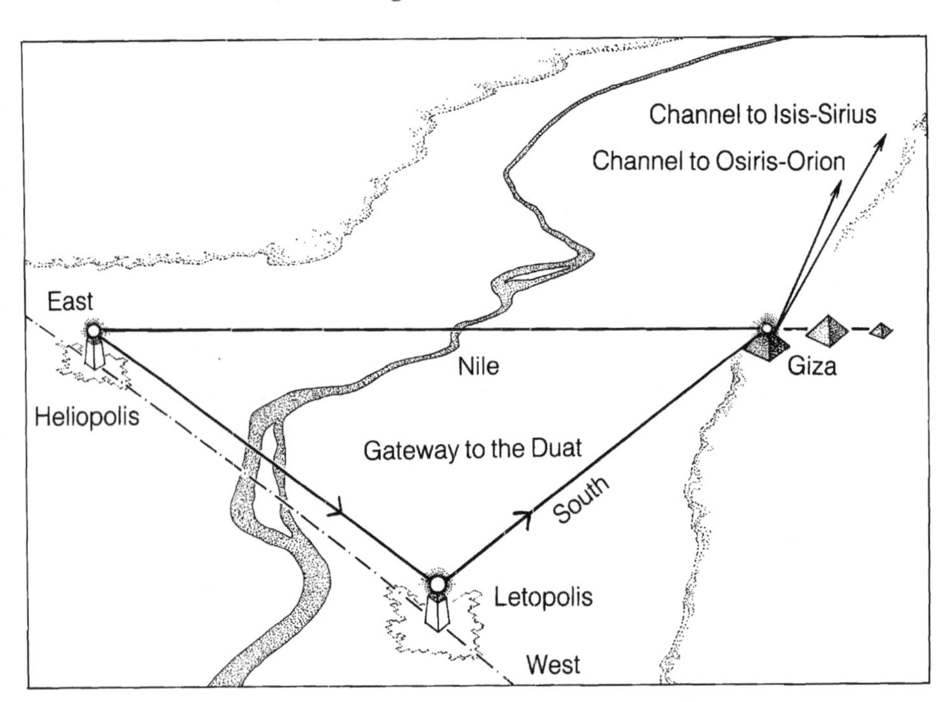

22. *Geodesic system linking Benben 'Beacons' at Heliopolis, Letopolis and Giza and final route of funeral procession*

Both Letopolis and Heliopolis are mentioned many times in the Pyramid Texts and were important religious centres in the Pyramid Age. Seen together, Letopolis and Heliopolis were aligned along a latitude and straddled the river Nile.[12] In the so-called Book of the Two Ways, written on the inside of coffins of the Middle Kingdom, El Bersheh,[13] we are told, 'I have travelled by the roads of Rostau on water and on land

. . . these are the roads of Osiris; they are [also] in the sky . . .'.

It is clear that the roads of Rostau (Giza) were across water and then on land, two major geodetic arteries or ways. This seems to define a religious procession starting from Heliopolis and travelling due west, across the Nile to Letopolis, then due south on land to Giza, ancient Rostau. We may thus suppose that before Giza there was a gateway into the Necropolis proper, symbolising the Gate of the Duat. We may also conjecture that the region which encompassed the cities of Heliopolis, Letopolis, Memphis and the pyramid region was a vast sacred site, a symbolic landscape with its counterpart in the sky near Sirius, Orion and the Hyades, along the banks of the Milky Way. We are satisfied that the case has been substantiated as far as present evidence allows, but there are these two major sites, Heliopolis and Letopolis, to account for. These cities also played a crucial part in the royal rebirth rituals of the Pyramid Age, for at Heliopolis was the Benben Stone, symbol of Osirian rebirth, and at Letopolis was the Horus of Letopolis priest responsible for the opening of the mouth of the Osirianised-king, and where the sacred adze instruments of *bja* were kept.[14] Where do these locations fit into the sky correlation map?

Egyptologist Georges Goyon, in his book *Le Secret des Batisseurs des Grandes Pyramides: Kheops*, comments on the position and alignment of the Great Pyramid:

> The monument [was] placed under the stellar protection of the god Horus, lord of Khem (Letopolis) . . . In order to direct the monument towards the sacred city of Khem, the astronomers determined the north targeting the north star, the polar (Alpha Draconis) . . . The recent discovery on the principle of orientation is based on the fact that all Egyptian pyramids of the Old Kingdom are oriented so that their north coincides with a sacred site or another pyramid which belonged to a venerated ancestor . . . Cheops's pyramid [is aligned] on Khem (Letopolis-Aussim) . . .[15]

Goyon believed that all Egyptian pyramids of the Old Kingdom were linked to a geodesic system involving a meridional grid across the Memphis region. Although he emphasised the meridian looking north, this same line is the south meridian if you direct yourself 180 degrees the opposite way, and it is likely that both the southern and the northern star systems were used by the ancient builders to fix the monuments on a meridian.[16] This is seen in the southern and northern shaft systems in Cheops's pyramid, where the southern shafts were directed to Zeta Orionis and Sirius, and the northern to Alpha Draconis and the star Beta Ursa Minor (Kochab) in the head of this constellation.

Goyon visualised the meridional link between the Great Pyramid and the city of Letopolis in an ingenious way. He lived in Egypt for many years and was the Egyptologist to King Farouk I;[17] he spent much of his time investigating the Memphis-Heliopolis-Letopolis region, and felt it necessary to ask:

> Did the Egyptians of the Pyramid Age already have astronomical and geodetic knowledge more advanced than we accord them? Did they know the geography of their country much better than we think? Had they already, in the third millennium BC, measured and gridded their land, in a manner claimed later by the Greek philosopher-mathematicians such as Thales, Pythagoras, Eudoxis, Plato, Democratis . . .?[18]

According to Goyon, the Greek geographer Strabo[19] said there was a great observatory near Letopolis called Kerkasore, which is also reported by Herodotus,[20] who says that Eudoxis and Plato made observations there.[21] Goyon asks whether there was not in the Pyramid Age 'another cause, an order of geodesy and mathematics?'[22] Much suggests that there was, and that the original geodetic centres were Heliopolis and Letopolis, which established a basic latitude and meridian. It was on this meridian that the unknown astronomer-priest, probably Imhotep as Chief of the Observers, fixed the position of the future Great Pyramid, the work of which began in the reign of Cheops (Khufu).

The correlation map of the terrestrial and celestial Duats of the Pyramid Age was established when the full sky-images of the risen Osiris-Orion and Isis-Sirius were seen over the eastern horizon: the moment when the sun was rising on the day of the heliacal rising of Sirius and near the summer solstice. Looking more closely at this sky-image, as reconstructed by the Skyglobe computer program, we see that the rising point of Sirius is about 26.5 degrees south of east and that the sunrise point is about 26.5 degrees north of east. Sirius lies almost directly below Orion's Belt and more precisely Zeta Orionis, which corresponds in the correlation map with the Great Pyramid. The horizon thus links the sunrise point and the star Sirius, sweeping a long line which divides the visible world and the invisible world beneath the horizon. At this point the sun is on the left side of the Milky Way, and Sirius, directly opposite, is on the right side, so the line between them has to cross the celestial river.

As we discussed in Chapter one, Heliopolis was the sun city par excellence, on the east bank of the Nile, and the city of Letopolis, on the west bank, is opposite Heliopolis.[23] Goyon confirmed that there seem to have been two high points, or mounds, one at Heliopolis and the other at Letopolis, from which the geographers made their geodetic sightings by observing gilded discs on top of pillars or obelisk-like monuments.[24] It is likely, however, that the gilded object at Heliopolis was not a disc but a pyramidion, probably the Benben itself gilded with gold-leaf and put (as Frankfort and Mercer[25] believed) on the pillar of Heliopolis, which originally belonged to Atum.[26]

A fairly implicit text from the Middle Kingdom, now in the Louvre Museum,[27] addresses Osiris:

> Hail Osiris, son of Nut [sky goddess] . . . whose awe Atum set in the heart of men, gods, spirits and the dead; to whom rulership was given in Heliopolis; great of presence in Djedu [the Osirian pillar[28]]; lord of fear in Two-Mounds; great of terror in Rostau [Giza] . . . such is Osiris, king of gods, great power of heaven, ruler of the living, king of

> those beyond [the horizon] . . . who owns the choice cuts in House-on-High, for whom sacrifice is made at Memphis . . .[29]

An alignment link between the mound of Heliopolis and that of Letopolis, using gilded reflectors such as Goyon described, establishes the horizon of a terrestrial Egypt (the terrestrial Duat) as the specific latitude (east-west line) which links up Heliopolis, the Sun City, with Letopolis, the city of Horus, son of Isis and Osiris, and, in astral terms as the Pyramid Texts say, Horus who is in Sirius [PT 632]. Heliopolis is therefore positioned to mark the place of sunrise when transferred on the sky-correlation map, which is east of the Milky Way and its terrestrial counterpart, the river Nile. It can also be seen that Horus who is in Sothis, i.e., the stellar god of Letopolis, marks the position of the heliacal rising of the star Sirius. In this completed sky-correlation map we thus have the full expression of the Osirian Duat, not only in its visible form in the sky but of its 'time', denoted by the heliacal rising of Sirius and the rising sun near the summer solstice as they both align on the eastern horizon.

With this geodedic linkage or 'road' established between Heliopolis and Letopolis, the great funerary procession could then proceed from the 'Sun City' to Letopolis and collect the 'Horus' and his 'four sons'. 'Horus' brought along his magical adze and his 'four sons' probably acted as pallbearers for the coffin of the Osiris-king. In great pomp and grief the procession headed for Rostau (Giza), gateway to the Duat, the Osirian kingdom on earth and in the sky. We begin to see what was meant by Horus saying, 'I have travelled by the roads of Rostau on water and on land . . . these are the roads of Osiris; they are in the sky . . .'. In Rostau the coffin was placed in a temple, probably at the north entrance of the pyramid. Eventually the coffin, which may have resembled a golden form of Osiris,[30] was taken into the pyramid and probably placed in the rebirth or Queen's Chamber.

Judging from later drawings in the *Book of the Dead*, the mummy was then stood upright with its face towards the northern shaft of the chamber, perhaps representing the adze of Ursa Minor (though the shaft was of course sealed). It is also possible that the mummy was stored temporarily in the mysterious niche on the east wall of the chamber. Standing in front of the mummy was the Horus, carrying his adze, with its potent astral connotations, and leading his four sons and any other celebrants present. Then there was the ceremony of the opening of the mouth, giving new stellar life to the mummified king. If the opening of the mouth ritual did take place in the Queen's Chamber, it is probable that it was timed to coincide with when the star Kochab was aligned with the northern shaft of the chamber.

When the Osiris-Orion mummy was deemed to have been struck with the magical force that brought about astral rebirth, the star of the pharaoh was born. Since the ancient name of the Great Pyramid was 'the Horizon of Khufu', in astral terms this meant that the 'star of Khufu' would have to be reborn, i.e., to rise over the eastern horizon, and in *c.* 2450BC this actually happened. For as the tip of the celestial adze struck the meridian and aligned with the northern shaft of the Queen's Chamber, Khufu's star Alnitak (Zeta Orionis) appeared on the horizon! Osiris-Orion Khufu was indeed reborn as a star when the tip of the celestial adze struck midnight on the circumpolar meridional clock.[31]

As with the original Osiris, the last earthly duty of the reborn king was to seed the womb of Isis-Sothis and ensure a successor to the throne of Egypt. There may have been some sort of ritual enactment of the stellar copulation between Osiris-Orion and Isis-Sirius, as described in the Pyramid Texts [PT 632], may have involved the southern Sirius shaft of the Queen's Chamber.

His earthly duties completed, the Osiris-king (the mummy) was probably taken out of the Queen's Chamber, up through the Grand Gallery and into the King's Chamber. Another ceremony may have taken place here: the 'weighing of the heart' before the mummy was placed facing the chamber's southern shaft. Now came the great dramatic moment when the soul of the star king liberated itself from the material mummiform and

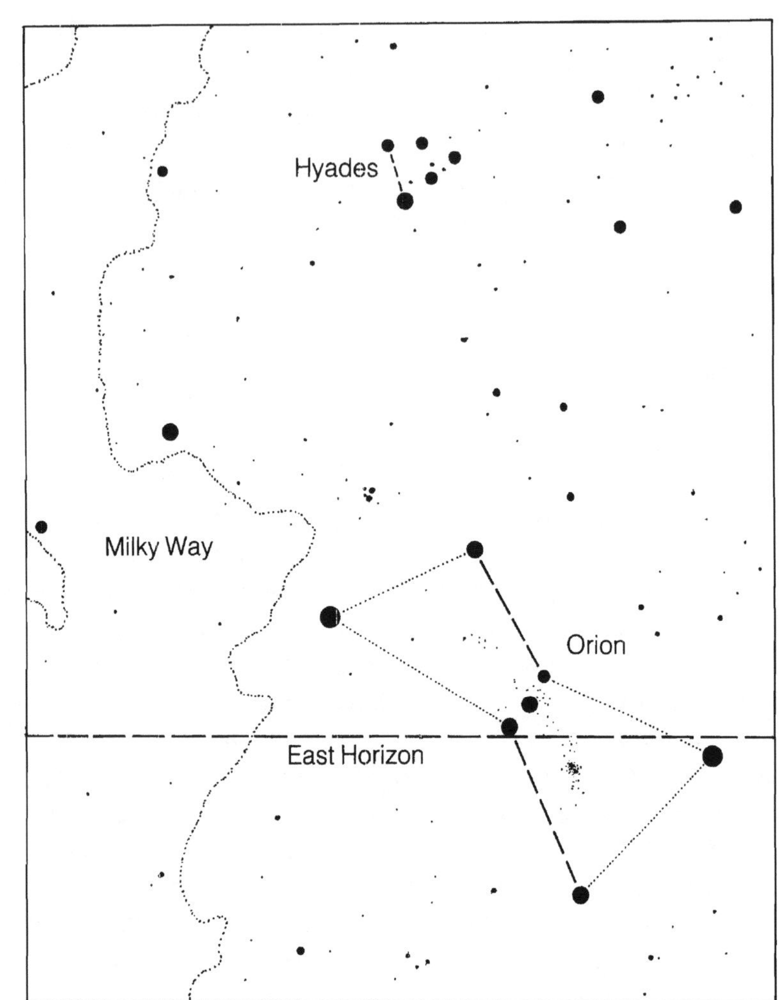

23. The Rising of Al Nitak c.2450BC

rose, through the southern shaft, towards the stars in Orion's Belt, the phallic region of Osiris-Orion in the sky. There the stellar king met the stellar form of his consort, Isis-Sirius, to create and give power to the new Horus-king, Horus who is in Sirius: 'Your sister (wife) Isis comes to you rejoicing for love of you. You have placed her on your phallus and your seed issues into her, she being ready as Sirius, and Horus-Sopd has come forth from you as Horus who is in Sirius.' [PT 632].

From this passage it is tempting to deduce that the southern shaft of the Queen's Chamber, targeted towards Sirius, served

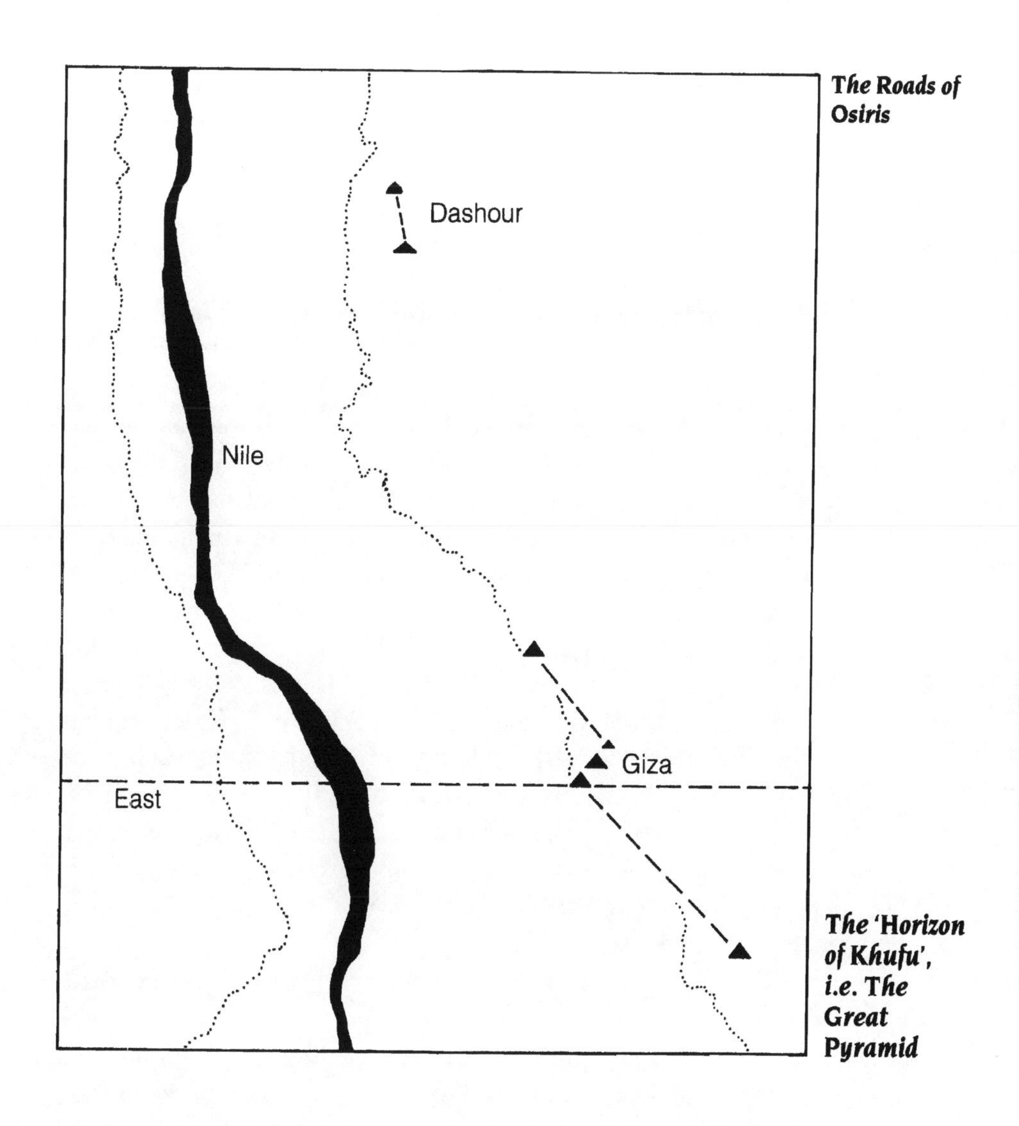

The Roads of Osiris

The 'Horizon of Khufu', i.e. The Great Pyramid

as a cosmic link between the phallus of the Osiris-king and the womb of Isis, (symbolised by the Queen's Chamber). There may therefore have been another ritual nine months later for the birth of the new Horus, some form of coronation ceremony confirming the new king as pharaoh of the two lands.

Viewed in this light, the Great Pyramid becomes the centre of the most important ceremonies of state and it is difficult to believe that it could have been used only once for the burial of Khufu and then sealed up for ever. While the presence of the granite plugs blocking the ascending gallery cannot be denied, we cannot be certain when it was that the pyramid was eventually sealed.[32]

Gantenbrink's remeasured angle of the southern (Sirius) shaft of the Queen's Chamber gave us the chance to confirm the symbolic archaeo-astronomical linkage between this shaft and the southern (Orion's Belt) shaft of the King's Chamber.[33] However, it should also be noted that there are physical links between the two southern shafts, for Gantenbrink has allowed us to reveal that directly above the place where the door is (at the end of the southern shaft of the Queen's Chamber) there is a small niche cut into the southern (Orion's Belt) shaft of the King's Chamber which passes directly above it.[34] This gives a geometrical, and probably a structural, link between the two shafts of the sort we expected to find as an outcome of the rituals described in the Pyramid Texts.

It should also be noted that the Queen's Chamber lies directly over the east-west axis of the pyramid, and thus on one axial line of the pyramid's capstone at the apex, where once stood a Benben or pyramidion.[35] The size of this Benben is not known, since it disappeared long ago.[36] Indeed, some researchers have suggested that it was not placed on the top of the Great Pyramid at all.[37]

The Great Pyramid is linked to Heliopolis by a geodetic system, so, symbolically, there was a signpost at Letopolis which linked the place of the Benben Stone at Heliopolis to the spot marking the centre-line of the Great Pyramid and thus the line through the two southern shafts towards the stars of the rebirth cult. Is this a clue that the end of the southern shaft of the Queen's Chamber may be linked in some way with the Benben Stone?

Since 22 March 1993 the world has been faced with the reality that there is a door at the end of this shaft. So a further question is: could the original Benben Stone be behind the door?

Further study of the Westcar Papyrus and illustrations from the Book of the Dead provide us with exciting possibilities. The

Westcar Papyrus tells us that Khufu was deeply interested in finding the secret number of the chambers of Thoth, supposedly kept in a shrine at Heliopolis, so that he could build the same for his pyramid.[38] The many illustrations of the opening of the mouth ceremony show the mummy standing with its back to a small shrine topped by a Benben. If we accept that the mummy is looking north (in the direction of the circumpolar constellations), represented, in these depictions, by the adzes of Horus and his four sons who stand in front of the mummy, the shrine must be to the south side of the rebirth room. In many of these illustrations, the shrine is shown to have a little door. We suspect that the southern shaft of the Queen's Chamber may lead to such a shrine. Now if we suppose that the Benben surmounting it is indeed the original Benben of Heliopolis, a further intriguing possibility presents itself. According to authors such as William Lethaby, the Benben of Heliopolis was itself a shrine,[39] believed to contain the lost books of Thoth, which, if they existed would have been written in the First Time, when Osiris was the ruler of Egypt. This again ties in with the prediction of Edgar Cayce[40] that, in the last years of the present century, a secret chamber containing records would be found in the pyramid.[41] If this prediction turns out to be true, we could be on the brink of finding the archetypes of the Pyramid Texts. The Great Pyramid might not, after all, be mute, as Mariette believed.

Finally, though, we have to ask: what if there is nothing at all and the mystery goes on? We will be content that, even if the Benben Stone and the shrine of Thoth are not at the end of this narrow shaft, we have, for the first time, discovered the true mystery of the pyramids: an earthly map of the stellar landscape of Orion the eternal home of the star-kings of Egypt.

EPILOGUE

Something else, however, was discovered inside the channels [of the Queen's Chamber] viz. a little bronze grapnel hook; a portion of cedar-like wood, which might have been its handle; and a grey-granite, or green-stone ball . . .'
– Charles Piazzi Smyth, *The Great Pyramid*: 1878

1. The Future of the Upuaut Project

In the long term, Rudolf's big hope for the future is to bring archaeology to the public in an exciting way and to raise global interest in the preservation of ancient sites around the world. With this in mind he has created The Upuaut Foundation in Monaco and is now in the process of putting together a specialised team of researchers and explorers. He has kindly asked me to be involved. In the short term, there still remains Gantenbrink's exploration of the northern shaft – possibly before Christmas 1993 – and, of course, the climax of his work when the little door is opened in the southern shaft in February or March 1994.

2. Mysterious Relics of Cheops

In early September 1993, that is nearly six months after Rudolf's discovery with UPUAUT 2 in the Great Pyramid (22 March 1993), I came across a rather startling passage in Charles Piazzi Smyth's book of 1878, *The Great Pyramid*, where I read an account of the 'newly discovered Air Channels in the Queen's Chamber'. Smyth described how Waynman Dixon and Dr Grant first discovered the shafts in this chamber:

> Perceiving a crack (first I am told, pointed out by Dr Grant) in the south wall of the Queen's Chamber, which allowed him at one place to push in a wire to a most unconscionable length, Mr W. Dixon set his carpenter man-of-all-works, by name Bill Grundy, to jump a hole with a hammer and steel chisel at that place . . .

Smyth then narrated how also the opening was found for the northern shaft and, too, how Dixon and Grant lit fires to check for outlets on the outside of the pyramid:

> Fires were then made inside the tubes or channels; but although at the southern one smoke went away, its exit was not discoverable on the outside of the pyramid . . .

But then followed a mysterious comment which, even more than a century later, made me jump out of my seat:

> Something else, however, was discovered inside the channels, viz. a little bronze grapnel hook; a portion of cedar-like wood, which might have been its handle; and a grey-granite, or green-stone ball . . . 8325 grains [about 0.850 kilograms] . . .

This was the very first time I had heard of this. I read on.

Smyth went on to explain how these relics or 'curiosities' had 'excited quite a furore of interest, for a time, in general antiquarian, and dilettante, circles in London; but nothing more has come of them'.

I found it odd that I had not heard about this before. My first reaction was to assume that the Egyptologists were well aware of the existence of such relics. I remembered the copper 'fittings' that Rudolf had discovered in the southern shaft. It appeared that he had not, after all, been the first to discover metal inside the Great Pyramid. I wondered what he would make of this. I immediately called Rudolf in Munich and, as I had anticipated, he was as astonished about this as I was. We both wondered why no Egyptologists had thought it important to inform us of Dixon's amazing find inside the channels in the Queen's Chamber. Perhaps they assumed that we already knew of this. I then called Dr I. E. S. Edwards but, to my greater surprise, he, too, had never heard of such items found by Dixon – nor had he come across Piazzi Smyth's report. He offered to check with the British Museum. It turned out that no one there could remember anything of this matter. Later Dr Spencer, who is responsible for the archives, also confirmed that no such items were recorded in the annals of the museum – let alone the relics themselves being there. This was most mysterious. What could have happened to these ancient relics from Cheops's pyramid? Were they brought to London after Piazzi Smyth examined them? From his account, this seemed to be the case.

It was then that I thought of calling an amateur astronomer I knew in Scotland. He put me in touch with Professor Hermann Brück and Mary Brück. Professor Brück had been Astronomer Royal for Scotland 1957–1975 and his wife was a lecturer in astronomy at Edinburgh University. They were the authors of several books, most recently a comprehensive biography of Piazzi Smyth. I telephoned Mary Brück and she told me she remembered seeing some drawings in Piazzi Smyth's personal diary of these relics. She kindly offered to research the matter. A few days later she reported that she had found many interesting letters and notes, and suggested

I come to Edinburgh. Two weeks later I drove to see the Brücks in their lovely home in Penicuik, near Edinburgh. To my great delight, Mary Brück produced a copy of Piazzi Smyth's diagrams showing the ancient relics and also, more interestingly, the various written accounts on them by the two Dixon brothers, Waynman and John.[1] From the accounts of Piazzi Smyth and the Dixons I felt that there were good chances that the relics might be found somewhere in London.

3. Secret Chamber Fever

The Dixon brothers seem to have been deeply involved with Piazzi Smyth since at least 1871. They, too, sensed the possibility of a 'secret chamber' in Cheops's pyramid. On 25 November 1871, for example, John Dixon reported to Piazzi Smyth that his younger brother, Waynman, was very busy working on a bridge construction project in Egypt and made this mysterious comment:

> I am more than ever convinced of the probability of the existence of a passage and probably a chamber containing possibly the records of the ancient founders – as soon as I have a decent plan drawn I will send you a copy . . .

John Dixon went to Egypt and when he returned on 8 April 1872, wrote again to Smyth, saying that Waynman was still very busy, and that 'I am satisfied I am on the clue to another passage!'

On 2 September 1872 a letter was written by John Dixon in London to Piazzi Smyth:

> I am gratified that our borings and scratchings at the pyramid have resulted in an interesting discovery of passages closely approaching the Queen's Chamber – I see he (Waynman) has sent you a copy of his

> report. I am anxious to have more by Monday's mail and shall send you a copy of his letter if he has not done so direct. I think the blocked entrance to them [the shafts] rather upsets the theory (?). I have further suggested to drill the west walls of both Chambers i.e. the King's and Queen's, also to see if by smoke and firing pistol in the passages they can by sight, sound or smell detect any connection with those of the King's. Possibly too the concussion may bring down any articles that have taken the benefit of . . . [the] 'angle of rest' and are lying up in the passages . . .

Then on 15 November 1872, John Dixon wrote a letter to Piazzi Smyth and mentioned again the 'Dixons Passages':

> I've just got back from a hurried visit to Egypt – seen the new passages or channels in the Queen's Chamber (Dixons Passages) – brought home the tools found in one – a bronze hook a granite ball doubtless a weight weighing 1lb 3oz – and a piece of old cubit five inches long . . .

4. The Missing Cigar Box

A few days later, on 23 November 1872, two letters followed from John Dixon to Piazzi Smyth. In one letter Dixon informed Smyth that he had dispatched the relics to him:

> These relics are packed in a cigar box and carried by passenger train. They consist of Stone Ball, Bronze Hook and Wood secured in glass tube . . . copy, photo or anything you like with them . . . but return them without delay as many are calling to see them and when next week *The Graphic* has a drawing of these in . . . there will be a rush . . . Is there any chance the British Museum giving a few hundred for these relics? If so, I'd spend the money in a great clearance and exploration [of the Pyramid

> base] . . . I'll beg them after their existence [the relics] become known . . .

In the second letter Dixon discussed Smyth's 'theory' that these shafts in the Queen's Chamber might have been 'air channels':

> Your remark as to the terminology of the new channels is forceful and good but I dissent from adopting on too hasty an assumption the theory that they are air channels for the obvious reason that they have been so carefully formed up to but not into the chamber. That 5 inches of so carefully left stone is the stumbling block to such a supposition. And again, one at any rate of them I am convinced from its appearance – so clean and white as the day it was made – cannot have any connection with the external atmosphere. It was here (in the north passage) we found the tools . . .

The now famous cigar box with the relics inside arrived safely on 26 November 1872 in the hands of Piazzi Smyth in Edinburgh. He entered this in his diary and also produced a full-size sketch of the metal 'tool'. Piazzi Smyth also correctly noted that the 'tool' was '. . . strangely small and delicate for [being a] Great Pyramid implement . . .'

On the 4 October 1993 I went to the Newspaper Library of the British Library at Colindale. I looked up the December 1872 issues of *The Graphic* and, in the issue 7 December 1872 I found John Dixon's article on p.530 (text) and p.545 (drawings).

From these, and Piazzi Smyth's own diagrams and commentaries of the relics, I concluded that the 'bronze tool' or 'grapnel hook' was an instrument used for a ritual, probably something to do with the 'opening of the mouth' ceremony. It reminded me of a snake's forked tongue. Such a 'snake-like' instrument was actually used in this ceremony and some good depictions can be seen in the famous Papyrus of Hunifer at the British Museum. The discovery of this implement inside the northern shaft, which we now know pointed to the circumpolar constellations – the sky region which is identified with this ceremony – adds further support to this thesis. Professor Z. Zäba, the astronomer and Egyptologist, has argued that an instrument called 'Pesh-en-kef', and shaped very much like the 'tool' found in the channel by Dixon, was, in actual fact, used in very ancient times in the ceremony of the 'opening of the mouth'. Furthermore, Zäba proved that the 'Pesh-en-kef' instrument, fixed on a wooden piece and in conjunction with a plumb-bob, was used to align the pyramid with the polar stars. It now seemed very likely that a priest placed the ritualistic tools inside the northern shaft from the other side of the wall of the Queen's Chamber.

Where could these relics be now? If not at the British Museum, then where? I took the diagrams of the relics to Dr Carol Andrews at the Egyptian Antiquities Department of the British Museum, but she seemed certain that they were not in their keep. Her first reaction was that the items, judging from the diagrams, did not look 'old enough', and she thought perhaps they were put in the shafts at a later date. But I reminded her that the shafts were closed from both ends until Waynman Dixon and Dr Grant opened them in 1872. The good state of preservation was actually explained by John Dixon in a letter dated 2 September 1872:

> The passage being hermetically sealed, there was no appearance of dust or smoke inside – but the walls were as clean as the day it was made . . .

Dixon was right, of course. With such a sealed system the relics were free from air corrosion. I gave Dr Andrews my opinion that the 'tool' was a Pesh-en-kef instrument, and also a sighting device for stellar alignments. Dr Andrews favoured the latter idea, but said that no Pesh-en-kef instrument of this shape was known before the Eighteenth Dynasty. I then showed the diagrams to Dr Edwards in Oxford and he, too, was compelled to support this idea but, unlike Dr Andrews, he recognised the instrument as a type of Pesh-en-kef. Both Rudolf Gantenbrink and I tend to agree with him on this.

5. Cleopatra's Needle and Victorian Memorabilia

The next place to check was at the Sir John Soanes Museum at Lincoln's Inn. John and Waynman Dixon seemed to know the curator, Dr Bunomi, at the time and so did Piazzi Smyth. But the archivist there, Mrs Parmer, was clear that no such items were ever given to the Museum. I told her of Bunomi's interest in Piazzi Smyth's theories and how he had been very excited by the arrival of Cleopatra's Needle in London. Apparently Dr Bunomi died in 1876, during the early stages of the operation to bring the obelisk from Alexandria. While we talked, Mrs Parmer remembered a curious event about Dr Bunomi: after his death, he had had placed on the roof of the museum a Doulton ware type jar full of curious memorabilia.

It was then that I suddenly remembered John Dixon's involvement with the Cleopatra's Needle affair. Both he and his brother, Waynman, had been contracted by Sir Erasmus Wilson and Sir James Alexander to supervise the transportation of the obelisk to London. But it was John who was primarily involved in the last stages of the operation and the erection of the monolith at the Victoria Embankment. The story appeared in the *Illustrated London News* of the 21 September 1878. I drove to the monument and read the commemoration inscriptions; one, on the north face of the monument, read:

> Through the Patriotic zeal of Erasmus Wilson, F.R.S., this obelisk was brought from Alexandria encased in an iron cylinder. It was abandoned during a storm in the Bay of Biscay, recovered and erected on this spot by John Dixon, C.E., in the 42nd year of Queen Victoria (1878).

According to the *Illustrated London News* of 21 September 1878, all sorts of curious memorabilia and relics were buried in the front part of the pedestal. These were put there by John Dixon himself in August 1878 during the construction of the pedestal, inside two Doulton ware jars. Among the strange items were 'photographs of twelve beautiful Englishwomen, a box of hairpins and other articles of feminine adornment . . . a box of cigars . . .'

Could John Dixon have put the ancient relics which he once kept in a 'box of cigars' under the London Obelisk? I telephoned an historian of the England National Heritage, Mr Roger Bowdler, but he did not think they had any details of the items

Entry 26 November 1872 from Piazzi Smyth's diary (by kind permission of Dr W. Duncan, Secretary to the Royal Society of Edinburgh)

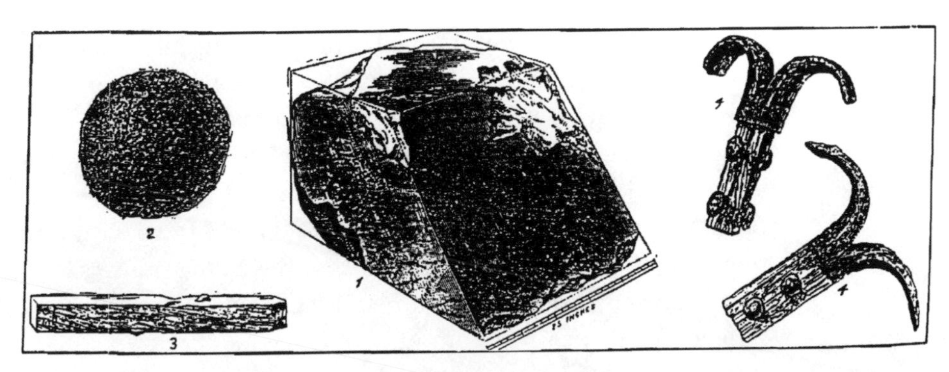

Discoveries in the Great Egyptian Pyramid
1. Original Casing Stone from North Side
2. Granite Ball, 1lb 3oz weight
3. Piece of Cedar, apparently a Measure
4. Bronze Instrument with portion of the wooden handle adhering to it.
The last three items were found in the northern shaft of the Queen's Chamber in 1872.

under the Obelisk. He suggested I try the Record Office of the Metropolitan Board of Works, who apparently were responsible for the operations to raise the obelisk in 1878. A frustrating search in the archives brought no result. Another search in the National Register of Archives also proved a dead end.

We cannot help wondering if these ancient relics – indeed, perhaps the very sighting instruments that were used to align the Great Pyramid to the stars – are in a cigar box under Cleopatra's Needle in London. Or perhaps they lie elsewhere, in some dark attic or cupboard in one of the many London antiquarian shops. We shall, perhaps, never know.

Postscript

It had been supposed that John Dixon and Piazzi Smyth erroneously described the 'tool' found in the northern shaft of the Queen's Chamber as being made of *bronze*. Egyptologists had always told us that the Bronze Age only occurred in Egypt the Middle Kingdom. Copper, therefore, was mistaken for bronze by the Victorians. To my surprise, on 2 November 1993, I was informed by Dr A. J. Spencer and Dr Andrews, both Assistant Keepers of the Department of Egyptian Antiquities at the British Museum, that two vessels of the Second Dynasty, previously thought to be copper, were now confirmed to be made of bronze. This meant that the description given by Dixon and Piazzi Smyth was, after all, correct! It also means that the Bronze Age had already started in Egypt centuries before everyone had assumed.

It was at that time that Dr Spencer also kindly allowed me to photograph an iron plate found in 1837 by a British engineer, J. R. Hill, stuck within a joint *inside the southern shaft* of the King's Chamber of Cheops's pyramid. It had been necessary to 'remove [it] by blasting the two outer tiers of the stones of the present surface of the pyramid'. Mr Hill and others with him then presented certificates stating that the iron plate was *contemporaneous with the pyramid*, and then deposited the ancient relic at the British Museum.[2]

The iron plate measures 26cm by 8.6cm. In 1926 Dr A. Lucas, the director of the chemical department at the Department of Antiquities in Egypt, examined it and 'thought that the iron was contemporaneous with the pyramid'; strangely, when he was told that it was not meteoric iron, he felt compelled to change his mind.[3] The matter lay dormant for more than fifty years, until in 1989, two eminent metallurgists, Dr El Gayar of the faculty of Petroleum and Minerals at Suez and Dr M. P. Jones of Imperial College in London, jointly performed chemical and microscopic tests on the mysterious iron plate and, to the annoyance of the British Museum, concluded that 'the plate was incorporated within the Pyramid at the time that structure was built'.[4] Their chemical analysis also revealed mysterious *traces of gold* and they conjectured that the iron plate might have been covered with gold. They also concluded that the plate was originally 26cm × 26cm (oddly, 26cm is exactly half an ancient Egyptian royal cubit, the measurement known to have been used by the pyramid builders) and thus probably was used to cover the mouth of the southern shaft some few metres from the outer face of the monument. If the conclusions of El Gayar and Jones are accepted – and we see no serious objections to them so far – it means that the Iron Age too began many centuries before Egyptologists had thought!

As yet, the 'Dixon relics' have not been found. The mystery of the great Cheops continues.

Appendix 1

ASTRONOMICAL INVESTIGATION CONCERNING THE SO-CALLED AIR-SHAFTS OF CHEOPS'S PYRAMID

by Virginia Trimble

The pyramid of Cheops at Giza is unique among the monuments of Egypt in several ways. Not only is it the largest, best built, and most thoroughly surveyed of the pyramids, but it possesses several architectural features not found elsewhere. Among the most obvious of these are two shafts leading north and south out of the King's Chamber and slanting up to open on opposite faces of the monument. Although the northern shaft makes an average angle with the horizontal of about 31 degrees and the southern one an angle of 44.5 degrees, because the King's Chamber is located south of the vertical axis from the apex of the pyramid, the two shafts open nearly at the same height on the northern and southern faces.[1]

The purpose of these shafts has not been determined, but it has frequently been held that they were intended simply for ventilation, hence the name 'air-shafts'. In view, however, of the profoundly religious character of the pyramids themselves for the Ancient Egyptians it seems not unreasonable to look for some deeper meaning to the shafts. It is the purpose of this paper to consider briefly some of the evidence for the view that the shafts were intended as ways whereby the soul of the deceased king might ascend to the north circumpolar stars and to the constellation now known as Orion. Although similar shafts do not appear to exist elsewhere, there is ample evidence for the presence of slots and apertures intended to allow the soul of the deceased to pass through various walls.

Such apertures first appear in the Third Dynasty tomb of Djeser[2] and become a regular feature in the serdabs of the Fifth Dynasty mastaba tombs.[3]

A notable feature of the religion of early Egypt was the 'stellar destiny' of the soul, wherein it was thought that the soul of the dead king would rise to the circumpolar stars – 'The Indestructible Ones' or 'The Imperishables' to the Egyptians – in their eternal journey around the sky. It is believed that the stairways or ramps descending from the north in archaic mastaba tombs were intended to aid the soul in its ascent to these stars. That the north shaft of the pyramid might have served a similar purpose is made more probable by its inclination. The latitude of Giza is about 30 degrees north (29 degrees 58 minutes 51 seconds), and we recall that the north shaft makes an angle of 31 with the horizontal.

This means that the shaft points very nearly toward the north celestial pole, about which the circumpolar stars seem to revolve. It is also of interest to note that, at the time the pyramid was built, the pole was marked by a bright star about as accurately as Polaris (alpha Ursae Minoris) now marks it.

It is generally known that the inclination of the earth's axis of rotation to the plane of its orbit (ecliptic) at an angle of about 23.5 degrees combines with the nonspherical shape of the earth and the gravitational force of the sun, moon and planets to produce a phenomenon known as the precession of the equinoxes. The effect of the sun and moon is to change the direction to which the earth's axis of rotation points relative to the fixed stars, while that of the planets is to change the plane of the earth's orbit relative to these stars. These effects are known as lunisolar precession and planetary precession respectively. It is evident that both factors will change the identity and positions of stars visible from a given point on the earth and that we must take them both into account when determining how the sky looked to the ancients.

In this scheme of moving stars, pole stars are a rather rare occurrence. In fact, after Polaris ceases to mark the pole in a few hundred years, there will not be another good one until alpha

Draconis returns around AD23000.[4] It happens, however, that the last 'visit' of alpha Draconis to the neighbourhood of the pole occurred from about 3000 to 2500BC.[5] This means that the Egyptians of the pyramid age were more aware than might otherwise have been the case of the apparent daily journey of the stars about a fixed point in the sky. It thus seems highly probable that they would have chosen to build a shaft that would allow the soul of their dead king to ascend directly to this central point.

Non-circumpolar stars were also of considerable importance to the Egyptians. They measured time at night by means of decans–stars or groups of stars which rose or culminated (reached their highest elevation above the southern horizon) at one-hour intervals during the night. Many of these decans were parts of constellation pictures (though different from ours which are derived from the Babylonian ones) and were identified with various gods. Very few of these have been identified with particular stars with any degree of certainty. There are, however, four of the standard decans and five variants thereof which are parts of the constellation *Sah* – 'The god who crosses the sky' – whose identification with Orion 'must be taken as likely in the highest degree'.[6] He is depicted as a man standing, looking back over his shoulder and holding a sceptre in one hand and an *'nh* (ankh) sign in the other. One of the five variants is probably the 'belt' of *Sah*. Three of the decans intended for use during the epagomenal days appear also to have been parts of this constellation.[7] We may note as evidence for the identification the ceiling of the tomb of Senmut, in which the column devoted to *Sah* includes three large stars arranged vertically and bearing a striking resemblance to the three stars we call Orion's Belt (delta, eta and zeta Orionis) which they probably represent.

The next relevant question is, of course, the position of these stars relative to the southern shaft at the time the pyramid was built. This requires calculations to allow for the two types of precession previously noted. We observe first that, because the shaft is directed due south, it can only point to a star at culmination, and we see that for a latitude 30 degrees north and the inclination of the shaft, 44.5 degrees, an appropriate star must have a declination (angular distance from the celestial equator) of −15.5 degrees. The question is then reduced to whether or not the stars of Orion ever had such a declination and, if so, when.

It can be shown by spherical trigonometry that, for a star at declination δ and right ascension θ (angular distance from the vernal equinox measured eastward along the celestial equator), precession will cause a change in position such that the declination at another time is given by:

$$\sin \delta' = \cos \delta \cos a \sin \theta + \sin \delta \cos \theta$$

where $a = \alpha + \equiv$, and θ and $\equiv$ are determined by the distance the ecliptic pole has moved due to planetary precession and the distance the north celestial pole has moved due to luni-solar precession during the given time. The values of these angles can be determined from the known rates and directions of the poles' motions. They have been tabulated for hundred-year intervals from 4000BC to AD3000 (for equinox 1900) by Paul Neugebauer[9] who has also worked out the right ascensions and declinations for 310 bright stars at hundred-year intervals from 4000BC to AD1900.[10] His tables and recent calculation by the same method show that one of the three stars in Orion's Belt had a declination within 30 degrees of −15.5 degrees (2840 to 2480BC). The positions of the stars during this period were:

Date	*Delta Orionis Declination*	*Epsilon Orionis Declination*	*Zeta Orionis Declination*
3000BC	−16° 51′		
2900	−16° 20′	−16° 47′	
2800	−15° 49′	−16° 17′	−16° 33′
2700	−15° 17′	−15° 46′	−16° 05′
2600	−14° 45′	−15° 16′	−15° 33′
2500	−14° 17′	−14° 46′	−15° 04′
2400		−14° 16′	−14° 34′
2300			−14° 06′

This means that these three stars, whose importance to the Egyptians we have seen, passed once each day, at culmination directly over the southern shaft of the Great Pyramid at the time it was built.[11]

Thus considerations of Egyptian religion and modern astronomy combine to indicate that the 'air-shafts' of Cheops's Pyramid were actually intended as ways by which the soul of the deceased king might ascend to join the circumpolar stars and the god constellation *Sah*.

It would seem likely that some other stars might pass in the same fashion over the opening of the shaft. It happens, however, that no other stars of comparable magnitude had declinations within 1 degree 30 minutes of −14 degrees 30 minutes during that period.

Originally printed in *Mitteilungen des Instituts für Orientforschung der Deutschen Akademie der Wissenschaften zu Berlin*, band x, Heft 2/3, 1964

Appendix 2

PRECESSION

by R. G. Bauval

Precession calculation is a vital tool for the historian to help him understand ancient man, whose religion was often directed to the 'sky gods' and thus based on observations of the sky – what today we would call naked-eye observational astronomy. It can be thus understood that ancient man built religious monuments, temples and, more prolifically, tombs which made use of geometrical astronomy to express astronomical alignments and other phenomena of the sky using symbolic architecture. It further follows that if an architectural feature of a monument is suspected to have been aligned towards a specific star, then, with the use of precession, it is possible to work out the date of such a monument to a fairly good level of precision. By also 're-creating' the sky for the given epoch, we can see what they saw and hence understand further the religious importance of their observations through the design and symbolic expression of the monument.

Before electronic scientific calculators and computers became household equipment, precessional calculations had to be done long-hand. These were not only complex but tedious, especially because the formulas combined spherical geometry and trigonometry through several steps of computations. If only one or two calculations were required, this was not so bad, but if several stars and dates had to be verified, the calculations might take all day. Happily for us today, a good personal computer does this for us: anyone with a PC can not only perform precessional calculation with a few touches of the keyboard but also actually see on the screen the effects

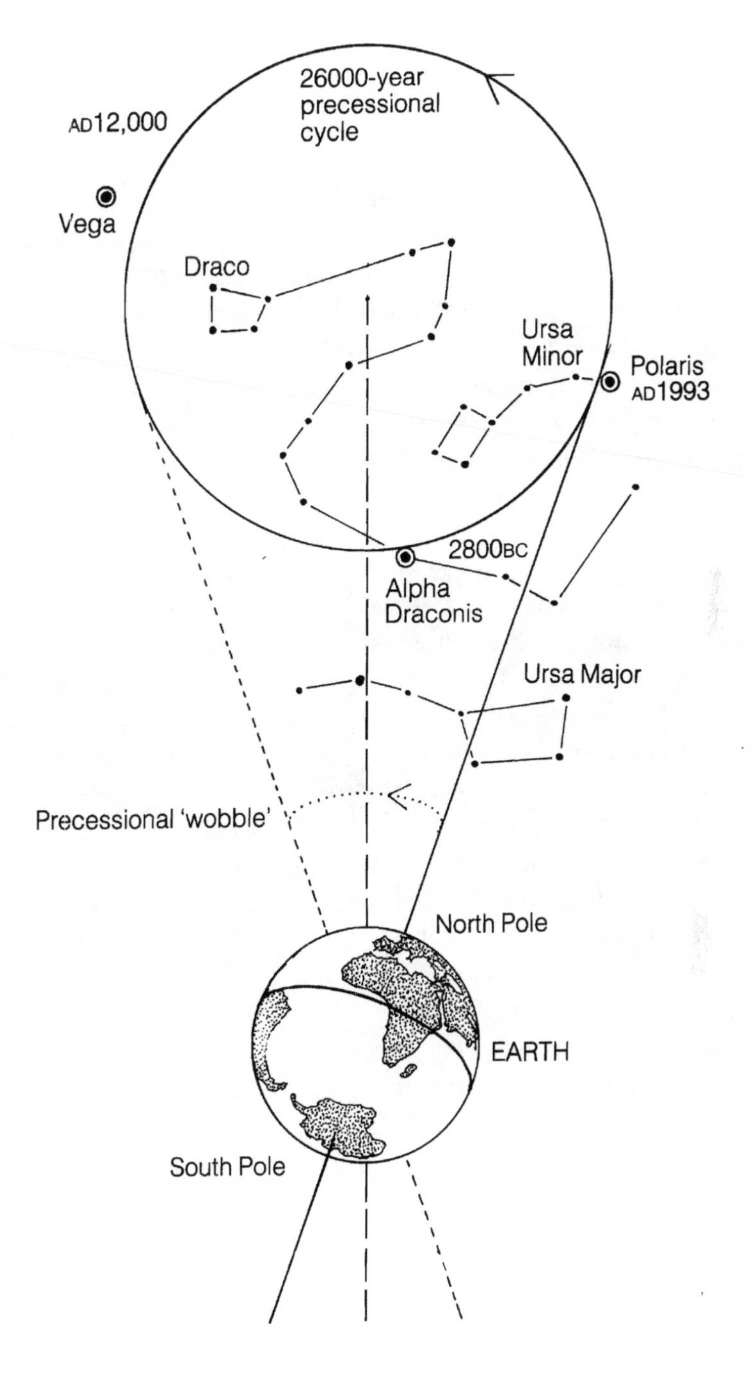

24. *Precession*

precession has on an artificial sky globe.[1] But what exactly is precession?

The sun and moon exert a gravitational pull on the earth's equatorial bulge, causing the planet to 'wobble' in a very slow cycle known as precession. The simplest way to think of precession is to imagine the earth as being a spinning top which also has a slow 'wobble' of just under 26,000 years' cycle. The extended axis through the poles thus performs a slow almost circular motion against the background of the starry sky and returns to the same place every 26,000 years. Every half-cycle of precession i.e. 13,000 years, a star finds itself in opposite direction on the precessional cycle, such that if it is observed at the high-point (highest declination) on the precessional cycle then 13,000 years later (or earlier) its position would be at its low-point (lowest declination) on the cycle.

The precessional effect is most noticeable at the meridian. Taking Orion's Belt as an example, in *c.* AD2550 it will be at highest declination (*c.* −0.8 degrees) very near the celestial equator. Thus it was at lowest declination (*c.* −48 degrees declination) in *c.* 10450BC. During the Pyramid Age, *c.* 2500BC, it was at *c.* −15 degrees declination.

The length of the precession cycle, however, is not absolutely constant but changes slightly from epoch to epoch. It is generally accepted, however, that it lies between 25,800 and 26,000 years. We have taken the value 26,000 years throughout *The Orion Mystery*. It must be noted that there is another shorter complex motion called nutation which takes 18.6 years.[2] This causes little 'hiccups' every 18.6 years on the otherwise smooth circular motion of precession. Nutation is generally ignored in precessional calculations for distant epochs since it is not possible to determine whether a hiccup was occurring at the date considered.

Both precession and nutation, of course, are not proper motions of the stars themselves, but are due to the movements of our own planet, producing the *apparent* motions of the stars. All stars, however, do have their own proper motions, i.e., they move in space. The closer the star, the greater visual effect its proper motion has over a given time. The farther the star, the smaller the visual effect. Proper motion is measured in angular change as a combination of declination and right ascension, these being the given co-ordinates of the stars on a sky map. Sirius is among the closest stars to our planet, at about 8.4 light-years away. The angular change due to its proper motion is given as −1.21 arcseconds per year. Over thousands of years this is quite noticeable and thus must be taken into account when precessional calculations are made. On the other hand the stars of Orion's Belt are very far indeed, about 1400 light-years, and generally no proper motion is registered.[3] Some researchers prefer to allocate a very small proper motion if a distant epoch is considered, but the resulting effect comes to well below the one-arcminute level for the epoch of the Pyramid Age. This cannot be perceived with the naked eye and thus proper motion is assumed to be negligible in such a case.[4]

When considering precession for relatively short periods of time, say fifty to one hundred years, the first approach is a simple rule of thumb where the sun appears to move against the background of the stars near the ecliptic (the path of the sun) by about 50.3 arcseconds per year. For 100 years this is about 1 degree 23 minutes and very noticeable indeed to a keen observer. Not all stars, however, are near the ecliptic and this rule of thumb cannot be simply applied to them. Nor does it show the effect of precession on declination. Mathematically, this is obtained by using the formula:

Change in Right Ascension (RA) = 3.07″ + 1.34″ sinRA tand then, Change in declination (d) = 20.0″ cosRA

In the case of very long periods of time, however, such as several millennia, a much more rigorous approach must be taken. In Sky Catalogue 2000.0, vol. I, the Rigorous Formula for Precession is given. Three auxiliary constants, A, B and C, are determined by the selection of the dates of the initial epoch (taken as AD2000) and the final epoch considered. These are given as:

A = 2305.647″ T + 0.302″ T2 + 0.018″ T3

$$B = A + 0.791'' T2$$

$$C = 2003.829'' T - 0.426'' T2 - 0.042'' T3$$

The first thing to do is to correct the position of the star for proper motion, given as (*u*)*RA* and (*u*)*d* for Right Ascension and declination respectively for one year, where the values of *u* are in arcseconds. This is done by multiplying (*u*)*RA* and (*u*)*d* by the number of years. The values are negative if before AD2000 and positive if after AD2000. The value (*u*) is the proper motion taken from tables. The result is added (forward in time) or subtracted (backward in time) from the Right Ascension and declination of the selected star's co-ordinates at the start of epoch AD2000. The new declination is given as *d* (*o*) and the new RA is given as *RA*(*o*). This therefore accounts for proper motion. The Rigorous Formula for Precession is then applied as follows:

$$\cos d(RA - B) = \cos do \sin[RA(o) + A]$$

$$\cos d \cos(RA - B) = \cos C \cos do \cos[RA(o) + A] - \sin C \sin d(o)$$

$$\sin d = \cos C \sin d(o) + \sin C \cos d(o) \cos[RA(o) + A]$$

A good scientific pocket calculator will do these operations quite easily. We have seen that there are other corrections than proper motion to be considered, such as nutation, and visual aberrations such as stellar parallax and refraction of light through the layers of gases in the atmosphere, but these are generally ignored. Allowing for their assumed effect may actually distort rather than improve the result by not knowing the exact value to consider, i.e., we have no way of knowing what was the density and clarity of the atmosphere on a given day in a given epoch. It is thus generally acceptable to ignore these effects, and assume that the plus and minus effects of nutation, aberration, parallax and refraction more or less cancel each other out.

Calculations made for me in 1987 by astronomer Dr John

O'Byrne of the University of Sydney revealed that for the three stars in Orion's Belt – Zeta (Al Nitak), Epsilon (Al Nilam) and Delta (Al Mintaka), no proper motion correction was considered necessary for the epoch 2500BC. Even by assuming a small value for proper motion effect, the correction needed would be about 65 arcseconds, which Dr O'Byrne felt would be 'unrealistically large'. Short-term effects such as nutation and aberration were ignored for the reasons given above.

For the star Sirius, a proper-motion adjustment of −1.21 arcseconds per year was required for declination. Going in negative time to the epoch of the Pyramid Age, this meant an adjustment of about +1 degree 33 minutes for epochs around 2500BC to precession had to be made.

For the purpose of *The Orion Mystery* we have used the Skyglobe version 3.5. This program has the advantage of providing very quickly a visual effect of precession and readings on the screen which give the declination, right ascension, azimuth, altitude and magnitude of a given star for a range of epochs for plus or minus 13,000 years. We found Skyglobe to be a very well-made program and quite accurate for the work covered in *The Orion Mystery*. Its accuracy is also very acceptable for the discussions. The star's co-ordinates, however, must be manually adjusted for proper motion. This was generally necessary for Sirius, whose proper motion is significant. We have, however, put the letter *c.* (*circa*) before dates signifying 'approximate'. In principle, precessional calculations dictate that the farther away the epoch under consideration, the greater the margin of error is for proper motion adjustments. No doubt professional astronomers, with more powerful means at their disposal, will find some hairs to split in the data provided in *The Orion Mystery*. Any refinement would, of course, be welcome. It must always be remembered that observations, for Ancient Egyptians, were made with the unaided eye and with the help of very basic sighting instruments. Values below the 20 arcminute level are not easily perceived with the naked eye. It is widely accepted that the Ancient Egyptians used a sighting instrument they called *Maskhet*: this was a wooden staff with a slit at one end, the latter used as a collimator to aim at stars.

They also used a simple plumb-line to measure the vertical.[5] With such sighting rods and plumb-lines, the altitude of a star at the meridian, or its azimuth at rising, can be measured with a very good degree of accuracy, certainly within the 20 arcminute level. Could the Ancient Egyptians have measured precession?

We have seen that precessional shift for, say, Zeta Orionis, which was then some 15 degrees south of the celestial equator, varied as much as 28 arcminutes in one century – equal to the apparent size of the moon. It is generally accepted by Egyptologists that the formative years or religious ideas predate the Pyramid Age by at least 500 years, and possibly much more. Thus over 500 years of observations, a variation of declination for Zeta Orionis between 2950BC and 2450BC would have registered about 2 degrees 16 minutes. This gives a rate of about 27 minutes per century for the change in declination. Having noticed that precession provided a uniform motion 'eastwards' of the sun along the ecliptic of some 1 degree 23 minutes per century relative to a given constellation or star,[6] it was not difficult for the Ancient Egyptians to deduce that a full cycle would take about 26,000 years to return to the same place relative to the constellation or star. Whether they worked out this value is debatable: what is more likely is that they realised that precession was a *cycle* (it has a start and an end), and then repeated the cycle for ever.

It is not known exactly when the Ancient Egyptians had developed a calendar, but it is generally accepted that this may have occurred well before the Pyramid Age.[7] In the calendar system used by the Egyptians, the year was divided into 12 months each having three *decans* of 10 days, thus 30 days in the month and 36 decans in a year. This gives a year of 360 days to which 5 extra or epagomenal days were added; these were called 'the 5 days upon the year'. It was during the 5 epagomenal days that the 'neters' or gods were born, who included Osiris and Isis. We thus have a situation where a 360-day year is linked to a 365-day year by the gods. The difference was, to them, caused by the birth of the gods who were said to be the four children of Nut (the sky goddess) Osiris, Isis, Seth and Nephthys with the fifth god being

Horus, son of Osiris and Isis.[8]

In religious terminology, it was thus the gods who turned the 360-day year into a 365-day year. These gods, as we have seen, were of course the stars. In this respect we must consider the question whether the Ancient Egyptians divided the apparent circular motion of the sun's ecliptic path around the earth into 'degrees' and, if so, was the division 360 units. It is a fact that the Egyptians divided the year into 12 months each of 30 days, giving the numerical total 360 days. They also divided the sky into 36 'decans' each of 10 days, also giving the numerical total 360 days. This implies that they divided the ecliptic path of the sun into 360 units or 'degrees' to define a day. But the correct numerical division should be 365 units, which they also had computed by adding the 5 days upon the year.

Appendix 3

THE SECRET CHAMBERS OF THE SANCTUARY OF THOTH

by Alan H. Gardiner

On the last day of October [1925] Professor Adolf Erman, the pioneer of modern Egyptian philology, attained his seventieth birthday. His pupils in various lands are celebrating the occasion in a special number of the *Zeitschrift für ägyptische Sprache*, but as one whose debt to the German scholar is particularly great I desire also to pay him some tribute in my own country. Now it was the intensive study of one particular papyrus containing a series of stories supposed to be told to Cheops, the builder of the Great Pyramid, which contributed more than all else to consolidate the foundations of our present knowledge of the Egyptian language. Professor Erman tells us that his edition of the Westcar Papyrus took him five years; he even devoted a special volume to its grammar. It is astonishing how well the translation which he published in 1890 has stood the test of time; in only a few details have his renderings or readings been questioned, although our progress both in lexicography and in grammar has been gigantic. For this reason any advance in the interpretation of the Westcar Papyrus seems rather an event, seems to register a step forward more significantly than would the novel translation of a passage in any other papyrus. I think to have found the solution of an old *crux interpretum* in the Westcar Papyrus; this solution I offer for Professor Erman's consideration in token of much gratitude.

The stories told to Cheops by the three first princes, his sons, related to earlier times; the fourth son, Hardedef, now promises to bring before his father a living man able to perform the most miraculous feats. This was a certain Djedi, who in spite of his hundred and ten years enjoyed an enormous appetite, was able to replace a head that had been cut off, and had the power to compel a lion to walk tamely behind him. In addition to these accomplishments he knew the number of the *ỉpwt* and of the *wnt* of Thoth, for which Cheops had been long looking, in order to make the like thereof for his own 'horizon', that is to say, for his own tomb (7, 5-8). The nature of the *ỉpwt* and of the *wnt* mentioned in this passage presents a problem. The *wnt* is, from its determinative, a building or structure of some sort, and the resemblance of its name to the name of the city where Thoth was particularly worshipped, namely *Wnw* Hermopolis Magna, the modern Ashmunên, would seem to indicate that it was the primeval sanctuary of Thoth, or else his tomb. Professor Erman thought that the resemblance of *wnt* and *Wnw* was fortuitous; this is also a possibility, but in any case *wnt* seems likely to be some special building dedicated to Thoth. The Pharaoh is said to be seeking (*ḥḥy*), not the *wnt* of Thoth, but the *ỉpwt* of the *wnt* of Thoth, whence it has been concluded, partly on other grounds to be examined later, that the *ỉpwt* were no longer in their original *wnt*. This again is a possible view, but not a necessary one; since Cheops was anxious to make for his tomb something like the *ỉpwt* of the *wnt* of Thoth, it is not unnatural that the writer should have said that the king was searching for these, and not for the *wnt* itself. There is no definite ground, in the passage before us, for asserting that the *ỉpwt* had been removed from their original *wnt*. I have no light to throw on the whereabouts of the *wnt*; it may be the name of the sanctuary of Hermopolis Magna, or it may be the name of an earlier sanctuary of Thoth in the Delta; or again it may be a purely mythical building. But that it was a building consecrated to Thoth, and that the *ỉpwt* were its secret chambers and hence inseparable from it, I hope to be able to prove, or at least to make exceedingly probable.

In 7, 5.7 the word *ỉpwt* appears to be determined with the sign of the bow , but in 9,2 we find not (7,7) nor (7,5) but , with the determinative of the cylinder seal which serves (*inter alia*) to determine the word *ḫtm* 'to seal up' or 'close'. On the strength of this determinative Professor Erman concluded that *ỉpt* denoted a closed building or the instrument for closing a building (*den Verschluss eines Gebäudes*). Now the later passage mentioning *ỉpwt* (9,1-5) reads as follows: '*Then said king Cheops* (namely to Djedi): *What of the report, thou knowest the number of the* ỉpwt *of the* wnt *of Thoth? And Djedi said: So please thee, I know not the number thereof, O Sovereign my lord, but I know the place where . . .* . *And His Majesty said: Where is that? And Djedi said: There is a box of flint in a room called 'Revision'* *in Heliopolis; (well,) in that box!*' In the following sentences Djedi declares that it is not he who will bring the box (*ʿfdt*) to the Pharaoh, but the eldest of the children who are in the womb of Reddjedet. This leads on to the well-known episode of the birth of the triplets destined to become the founders of the Fifth Dynasty.

Now Professor Erman rendered the words omitted in the above translation as 'the place where they are', and it must be admitted that in the absence of any evidence as to the nature of the *ỉpwt*, this seems necessarily the right translation. Hence it was naturally concluded that the *ỉpwt* were small enough to be contained within a box, and no surprise was felt when Mr Crum subsequently produced a Coptic word ⲉⲡⲧⲱ in close association with other words for 'doors', 'bolts', 'keys' (*Zeitschr. f. äg. Spr.*, XXXVI, 147). Since that time *ỉpwt* has been translated 'locks', and it is supposed that Cheops was searching for the locks of the *wnt*-sanctuary of Thoth, and that Djedi declared these to be in a flint box in the temple of Heliopolis.[1]

In opposition to this theory it must be noted, first of all, that the rendering 'locks' rests wholly on the determinative … which *ipwt* has in 9,2 and nowhere else, either in the Westcar Papyrus or out of it; secondly, that the determinative … accords ill with the meaning 'locks';[2] and thirdly, that the determinative … found in the passages 7, 5.7 is left without explanation. It is evident to me that the hieratic sign transcribed … is really the equivalent of …, though the proof of this fact is a little roundabout. Möller cites no early equivalent of …, though I think that the obscure sign in *Sinuhe R7*[3] and another rather different form in *Sinuhe* B205 are examples from Twelfth Dynasty and rather later. From the Hyksos period, however, no instances are forthcoming unless it be the two in the Westcar Papyrus here cited. Now we have proof that in hieroglyphic of the New Kingdom … and … are constantly confounded (*Zeitschr. f. äg. Spr.*, XLV, 127), and in my *Notes on the Story of Sinuhe*, 152, I have quoted an autobiographical stela of about the reign of Tuthmosis III where … seems a pretty obvious quotation of *Sinuhe* R2-3 … *'He said: I was a follower who followed his lord, a servant of the royal harîm.'* The confusion of … and … must obviously be due to the similarity of these signs in hieratic, so that we may regard it as an acquired fact that before the reign of Tuthmosis III the hieratic forms of … and … looked very much alike. Now if the student will consult the *Carnarvon Tablet*, l.1, dating from at latest the beginning of the Eighteenth Dynasty,[4] he will there find … *nst* 'throne' written with a sign almost identical with … ; *nst* has a similar shape in *Sinuhe* B207. In view of these coincidences, it is impossible to doubt that … and … have to be read in *Westcar* 7, 5.7; in *Westcar* 9,2 … is merely an erroneous substitution for the rarer sign. Our translations of the passages in question have to be remodelled accordingly.

Apart from the *Westcar* passages and the name 'Southern

Opet' [hieroglyphs] given to Luxor, the word *ip{{CONTENT}}gt;t* is almost always used in reference to the royal harîm as a locality; see *Zeitschr. f. äg. Spr.*, XLV, 127. It seems likely that the word signified properly a secret or privy chamber. Applying this rendering in 7,5-8, we find that the delight of Cheops at the prospect of seeing Djedi was due to the fact that the latter '*knew the number of the secret chambers of the sanctuary of Thoth*', for Cheops himself '*had spent (much) time in searching for the secret chambers of the sanctuary of Thoth in order to make the like thereof for his horizon*'. And indeed, what ambition could have fired Cheops more than to possess in his own pyramid a replica of the mysterious chambers in the hoary sanctuary of the god of Wisdom? The temple of the Great Pyramid is utterly destroyed, but the inner chambers of the pyramid itself remain a marvel down to the present day. So much for the first passage; the second is a little more difficult to interpret. We have seen that the words [hieroglyphs] are most easily rendered '*(I know) the place where they are*', in which case, as the following question and answer reveal, the *ỉpwt* of the sanctuary of Thoth would be in a flint box in a room of the temple of Heliopolis. This view of the meaning is, of course, incompatible with the sense 'secret chambers' which we now attribute to *ỉpwt*. Let us re-examine the passage afresh, attempting a different translation. Cheops asks whether Djedi knows the number of the secret chambers of the sanctuary of Thoth. Djedi replies: *So please thee, I know not the number thereof, O Sovereign my lord, but I know the place where it* (scil. the number or the knowledge of the number) *is*. He then proceeds to say that '*there is a box of flint in a room in Heliopolis called "(the room of) Revision"; in that box (the information will be found)*.' According to this mode of understanding the passage, what was in the flint box is not the *ỉpwt*, the secret chambers themselves, but a papyrus recording their number. Objectors to this view can make some capital out of the fact that the text *bw nty st ỉm*, not *bw nty sw ỉm* with the masculine pronoun *sw* which would be expected if the reference were to *tnw* 'the number'. But possibly the vague neuter pronoun *st* 'it' may refer, not to the specific word *tnw* 'number', but to the required information generally. I admit there is some difficulty in taking this view, but an argument can now be adduced which makes it practically certain that this is the view to take. Insufficient weight has been attached to the name "Revision" [hieroglyphs] given to the room in which the flint box was to be found. Now *sipty* is the regular word employed for 'taking stock' of the property of a temple, as Professor Erman himself has shown.[5] For this reason, surely, the room in question must have been an archive, not a storehouse of any kind. I conclude, therefore, that the word *ipwt* means 'secret chambers', and that Cheops was seeking for details concerning the secret chambers of the primeval sanctuary of Thoth, in order that he might copy the same when building his pyramid.

This article appeared in the
Journal of Egyptian Archaeology, 11, 1925

Appendix 4

THE SURVIVAL OF THE STAR RELIGION

by Robert Bauval and Adrian Gilbert

Within the core-thesis of the Orion Mystery lies not only the correlation theory of the Duat with the Fourth Dynasty pyramid fields but the fact that the dominant religion of the pyramid builders was a star religion, and that the dead kings were supposed to become star souls of Orion. The question is whether the star religion of the Pyramid Age, so vividly expressed in the Pyramid Texts of the Fifth and Sixth Dynasties and in the archaeo-astronomical language of the great Fourth Dynasty, persisted through the whole pharaonic era – almost three millennia, from the first native dynasty in 3100BC to the end of the last in *c.* 525BC, and indeed beyond.

Ancient Egyptian chronology is something of a nightmare, with no two scholars agreeing on precise dates. This is especially so for the earlier dynasties where, according to experts, at least 150 years, plus or minus, must be assumed as the margin of error. But there is a consensus which we will adopt here. The Ancient Egyptians did not, of course, see themselves as dynasties but as a continuous line of kings which began in the First Time when the gods ruled Egypt. They saw Horus, son of Osiris and Isis, as an historical person who became the first man-god to rule Egypt as pharaoh. The term Pharaoh comes from *Per-Aa* and means 'Great House', the pantheon or great divine house from which the kings of Egypt came. All pharaohs saw themselves as the reincarnated Horus, the Living One as opposed to the dead and Reborn Ones who had departed into the astral afterworld and had themselves become an Osiris or star soul.

To conform to modern Egyptological practice, we shall assume the so-called dynastic divisions. Pharaonic Egypt, which lasted from about 3100BC to 332BC – and thus vastly longer than the Greek and Roman put together, and indeed Western civilisation as a whole – included thirty-one distinct dynasties with some 390 monarchs.[1] Although there were 'pharaohs' after 332BC until AD251, these were not native kings but Macedonian Greeks (Ptolemaic period 332–30BC) and later Roman emperors (Roman period 30BC to AD251). Including these, there were 439 monarchs who ruled Egypt as pharaohs.[2] Egyptologists have thought it best to separate such a lengthy epoch into periods, and these are shown in the Table.[3]

Dynasty	*Period*	*Years*
1–2	Early Dynastic	3100–2686BC
3–6	Old Kingdom	2686–2181BC
7–10	First Intermediate	2181–2133BC
11–12	Middle Kingdom	2133–1786BC
13–17	Second Intermediate	1786–1567BC
18–20	New Kingdom	1567–1080BC
21–25	Late New Kingdom	1080–664BC
26	Sait	664–525BC
27–31	Late	525–332BC
	Ptolemaic	332–30BC
	Roman	30BC–AD642
	Arab	AD642–present

From the evidence in the Pyramid Texts and the monuments themselves, it is clear that the rebirth cult was focused on the king alone or may, at most, have extended to members of the royal family. Only they had a right to an astral rebirth which involved mummification and the complex rituals performed, no doubt, in the pyramid zone of Memphis and even, we suspect, within the pyramid structures.

There can be little doubt that during the epoch of the great Fourth Dynasty the central point of the rebirth rites was Giza, and that the Great Pyramid serviced the apotheosis of an ancient passion play involving the body of the dead king and a royal and priestly congregation. The sharp decline at the close of that dynasty is evident from the smaller and poorly constructed pyramids of the Fifth Dynasty kings at Abusir and Saqqara. From this point, or at least from the end of the Old Kingdom, the royal rebirth cult became more democratised, extending to notables at the court and probably even rich merchants and military men. As the process expanded, it seems likely that more and more commoners were given the right to an astral rebirth, so that by the time of the New Kingdom everyone in Egypt who could afford the expenses of mummification, the elaborate funeral and accompanying paraphernalia, was allowed a life after death with Osiris. However, democratisation brought a gradual corruption of the cult and variation of the rituals to suit the special needs of the deceased and his favourite local gods. In short, the rebirth cult lost its purity and simplicity.

The textual route showing the survival of the stellar rebirth cult is mainly through the various versions of the Book of the Dead, of which the Pyramid Texts is the oldest version. There are also the many inscriptions found in tombs and temples and, of course, the large collections of papyri in museums around the world. A detailed study of all this material is well outside the scope of this book; what we can do is to draw on selected texts which leave no doubt that the Osirian afterlife prevailed throughout the pharaonic era and that the destiny and final form of the dead remained astral – a star soul in the Duat or afterworld kingdom of Osiris.

The persistence of the star religion in the Old Kingdom and Pyramid Age has been presented in this book by investigation of the Pyramid Texts. The next set of textual material to examine – the natural follow up to the Pyramid Texts – is the so-called Coffin Texts of the Middle Kingdom, the epoch which followed the Pyramid Age. Carol Andrews, a senior Egyptologist at the British Museum, says:

> The Middle Kingdom (about 2040–1786BC) was a time when funerary beliefs and practices were democratised, when a guaranteed afterlife, which before had been restricted to royalty and great noblemen, became available to all who could acquire the relevant equipment. Now to the Utterances of the Pyramid Texts were added many more spells, and this new repertoire was written not in hieroglyphs but in the cursive script called Hieratic, in closely crowded vertical columns within wooden coffins of commoners. Because of their new locations the spells are now known as the Coffin Texts, and it is they which are direct predecessors of texts written in Book of the Dead papyri of the New Kingdom and later.[4]

It is pretty clear that the Pyramid Texts were the predecessors of both the Coffin Texts and the Book of the Dead, which eventually takes us to the Ptolemaic period, the few centuries which predate the early Christian and Gnostic epoch. Carol Andrews goes on to say, 'A new development in the Coffin Texts is that the sun god is no longer supreme: Osiris is the king under whom the blessed dead hope to spend eternity, the god with whom the dead became assimilated . . .'[5]

Andrews also says that in the Coffin Texts a new concept appears: the afterlife is spent in the 'Fields of Reeds', where agricultural activities undertaken by the dead mirror the activities in Egypt, so that the 'Other World was envisaged as an identical environment'.[6] The Fields of Reeds is, however, not a new concept of the Coffin Texts but comes from the Pyramid Texts and hence the Pyramid Age. In Faulkner's edition of the Pyramid Texts the Fields of Reeds are mentioned many times in direct connection with the afterlife destiny and are obviously visualised as a celestial and thus astral landscape which resembles the Nile region of Lower Egypt and are an integral part of the Duat. I. E. S. Edwards says of the Fields Of Reeds: 'Even in earlier times, however, the Osirian hereafter was probably regarded as a kind of idealised version of this world, situated below the western [sic] horizon and presided over by Osiris. This region, called by the Egyptians the Fields of Reeds, was subsequently known to the Greeks as the Elysian Fields . . .'[7]

Edwards remarks emphatically that the Ancient Egyptians 'regarded the after-life as a kind of mirror of this world' and that it was a place where the dead 'spirits could thus dwell at will near Osiris'.

In the Coffin Texts the Nile god says 'I am he who performs the service of gifts (the harvest) for Osiris at the Great Inundation, I raise up my divine command at the rising of the Great God (Osiris).'[9]

Also in the Coffin Texts we read that 'Osiris appears whenever there is an outflow' of water, i.e., the annual flood.[10] 'The rising of the Great God' at the start of the Nile's flood offers us the imagery of the rising of the astral Osiris (Orion). Thus, in this spiritual or soul form, says Rundle Clark, 'Osiris is especially considered the spirit in the Nile flood . . . The rising of Orion in the southern sky after the time of its invisibility is the sign for the beginning of a new season of growth, the revival of nature. Osiris has been transformed into a "living soul" . . .' i.e., a Ba or star soul, in this case Orion. The idea that the Ba was indeed a star soul can be found throughout the pharaonic epoch, in the so-called Papyrus Carlsberg I, for instance, which dates from the second century AD, well into the Christian era. The Carlsberg I papyrus, now in the University of Copenhagen, came originally from the Fayum, a fertile oasis south of Cairo much frequented in the second century AD by Christian Gnostics.

Similar texts are known as the Dramatic Texts, and come from the tomb or cenotaph of Seti I at Abydos (*c.* 1350BC), where they still are. Otto Neugebauer and Richard Parker, experts in Egyptian astronomy, say that 'in chapter VI, 43, the souls are referred to as "stars" . . .'[11] The actual passage in the Dramatic Texts, Part II, VI, 43 which Neugebauer and Parker refer to, reads, 'The souls go forth and they travel in the sky at night. The rising of the stars. They travel at night . . .', and goes on, 'when it (the soul) is seen by the living, it is indeed a star, the people do not see it by day . . . One sees that is how it (the soul) lives there. You see it shining forth in the sky . . .'

The Carlsberg I papyrus, which draws much of its material from cosmology on the ceilings and walls of Seti I and the Ramesside tombs (*c.* 1300–1150BC), is a detailed treatise of the rebirth of human beings as stars in the Duat. A few quotes from the text and also commentaries from Otto Neugebauer and Richard Parker, who have studied it for many years, summarise the essential points:

> the most important information that comes from this chapter (Carlsberg, I, Ch. E) is the fact that the decans (groups of stars) indicate the hours no longer by their successive rising but by their culmination (at the meridian) or transit. The star of the 'first' hour is the decan which has completed its ten days as first hour star and is seen at the meridian at the beginning of the night, that is, sometime after sunset . . .[12]

The writers go on to explain that after this meridian passage, a star is seen to take ninety days (three months) to reach the western horizon at the same time of day (i.e., dusk, just after sunset). Then it 'enters' the Duat, that is, it becomes invisible for a period of seventy days. The seventy days, say Parker and Neugebauer, is modelled on the period of invisibility of Sirius. Then the star is reborn in the east; it 'comes forth from the Duat' and travels the sky from east to west. It takes eighty days to reach the meridian, this time at dawn, before sunrise: the twelve hours of this star. Another 120 days (twelve decan hours) sees the star at the meridian at dusk, just after sunset. This is its 'first' hour, and the cycle begins again. It appears that the star works (is an active soul) only when it can be seen to cross (transit) the meridian, eighty days after its rebirth, its helical rising, when it is at the meridian at dawn. A simple calculation thus shows that the decan or star works for 120 days, that is, twelve decan hours of ten days each.[13]

Also contained in these texts is the concept that rebirth of a soul star occurs at its heliacal rising; when it rises in the east at dawn after its seventy-day period of invisibility. The star is thus imagined as emerging from the female figure of the sky goddess as she is arched across the sky with her thighs in the east. The following passages are inscribed next to an image of the sky goddess arched in that position:

> The female figure of this (figure) . . . that is to say her head is in the west and her hind part in the east . . . He causes the hind part to be the beginning, that is to say, the Place of Birth . . .[14]

> the marshes of heaven of the gods (stars), is the place from which the birds (Ba-souls) come . . . they are from the north-west side . . . as far as the south-west side . . . of the [sky] . . . which opens to the Duat which is on the northern side [of the sky] . . .[15]

Clearly the dead person's soul yet to be reborn enters the Duat in the north or circumpolar region but then starts its labour (work), presumably of being gestated inside the womb of the sky goddess, when the star is at dawn at the meridian. It takes 90 + 120 + 70 = 280 days to complete its astral gestation and for the soul to be reborn at its heliacal rising in the east at dawn. The average time for human gestation is, of course, 280 days.

The Texts go on to tell us that the special stars under consideration rise in the south-east part of the horizon, where Orion and Sirius rose (and still do):

> . . . these are the risings of the gods. These . . . Orion and Sothis (Sirius), who are the first of the gods – that is to say they customarily spend seventy days in the Duat [and they rise] again . . . It is in the [south] east that they celebrate their first feast . . .[16]

Finally the Texts reveal that the life-death-rebirth cycle of a star is regarded as the same for humans:

> . . . their burials (the stars) take place like those of men . . . that is to say, they are the likeness of the burial-days which are for men today . . . seventy days which they pass in the embalming-house . . . Its duration in the Duat indeed takes place. It is the taking place of its duration in the Duat . . . every one of the stars – that is to say 70 days . . . this is what is done (meant) by dying. This one which sets is the one which does this . . . the star among them which goes to the Duat . . .[17]

The Neugebauer–Parker commentary on these texts is that the analogy of human embalming and 'the stay of a star in the Duat for seventy days' is made explicit. They go on to say, surprisingly, that 'no suggestion has yet been made why seventy days has been chosen for the ideal period'; then conclude, rightly, 'it is the behaviour of Sirius – the prototype of the decanal stars – which suggests it.'[18]

It is apparent that the event of a human death and rebirth in an afterlife world or 'cosmic Egypt' was based on the annual cycle of the stars and, more specifically, on those of Sirius and Orion, the divine couple and protagonists in the drama of astral rebirth. This idea has its origins in Egypt's early Pyramid Age, and was first expressed in the sacred astro-architectural language of the Fourth Dynasty, who built the Giza and Dashour giant pyramids. These pyramids have survived, together with the Pyramid Texts of the Fifth and Sixth Dynasties which provide us with the fundamentals of a potent star religion of rebirth.

This religion is the purest manifestation of the human hope that religious rituals and liturgies will assist the initiate or believer to achieve rebirth as a star soul in the afterworld of Osiris. When the texts are analysed as a whole, we conclude that a gestation cycle of some 273 to 280 days (about nine months) took place when a star began its labour at the meridian at dawn, to reach, in the east at dawn, the apotheosis of rebirth.

The cardinal points were of the utmost importance to the rituals involved: the south (meridian) marking the start of the cycle, the west the start of symbolic death when the star became invisible, the east denoting rebirth when the star rose heliacally. The north seems to have been regarded as a fixed point where the energy for the process could be generated, like a cosmic umbilical cord linked to the whole event. The mysterious abode of Tuart was there – the hippopotamus goddess of fecundity and childbearing, represented by the constellation we now call Draconis. Interestingly, the pole region of the sky Tuart inhabited also had a 'mooring post' from which a rope or cord emerges. This mooring post is often mentioned in the Pyramid Texts, in relation to the astral rituals, and is depicted in many astronomical drawings of a later period.

So, is the material in the Pyramid Texts, and its later version, the Coffin Texts and Book of the Dead, expressing the same thing as the astro-architectural language we read in the Fourth Dynasty pyramids, and particularly that of Cheops? We believe that the answer is 'yes'.

Let us go back to the myth of Osiris and Isis and take a closer look at it from an astral viewpont. Osiris was killed by his brother, Seth, and Isis gathered his scattered limbs and brought about his resurrection, but one vital part of his body was missing: the phallus. Isis had to use an artificial phallus to make herself pregnant and bring forth Horus. If we look at the Orion-Hyades star figure showing a male human shape, we can see how the region we call Orion's Belt fits the place of the phallus. It has often been suggested, (recently by the author), that the shafts in the Cheops pyramid served a fertility or phallic role in the stellar rebirth rituals.[19] It is therefore reasonable to assume that the three stars which form Orion's Belt represent the phallus of Sahu-Orion (Osiris-Orion). This has its counterpart on the ground in the three Giza pyramids, one of the southern shafts of Cheops's pyramid (the King's Chamber) being directed to Orion's Belt. The southern shaft of the Queen's Chamber was directed to Isis-Sirius, and this is found textually in the Pyramid Texts, where Osiris-Orion is addressed: 'Your sister Isis comes to you rejoicing for love of you. You have placed her on your phallus and your seed issues in her, she being ready as Sothis (Sirius), and Har-Sopt (the stellar Horus) has come forth from you as Horus Who Is In Sothis . . .' [PT 632-3].

We have reason to conjecture that we are being told of the shafts in the Cheops pyramid: that the phallus of Osiris-Orion is the southern shaft of the King's Chamber pointing to Orion's Belt, and is connected with Isis-Sothis (Sirius) through the southern shaft of the Queen's Chamber. The phrase 'Your sister Isis comes to you' indicates that there should be a physical link between the two shafts, and Gantenbrink may have found this link when he sent his UPUAUT robot up these shafts. At the end of the southern shaft from the Queen's Chamber (still with about nineteen metres to run before it could pierce the face of the pyramid) he found the little portcullis door. Directly above this point is the southern shaft of the King's Chamber, the Orion shaft, and here there is a marking or niche, indicating that the ancient builders saw a link between the two southern shafts.

If this conclusion is correct, this would force us to deduce that the large area between the two southern shafts may contain something to do with the stellar ritual for the seeding of Isis to create a symbolical new Horus-king to replace the departed king. This would be in line with the religious beliefs of the epoch, as the Pyramid Texts show. British Egyptologist Henry Frankfort, ex-director of the Warburg Institute in London, brought to light what he saw as a double event put into motion after a king died: the first was the funerary rites involving the elaborate preparation of the dead king as a Sahu (mummy or spiritual body),[20] which took the corpse to the brink of an astral rebirth; the other event, in parallel to the first, was the transfer of kingship to the new, living Horus-king.[21] (In May 1993 Robert Bauval was invited by Dr Nicolas Mann, director of the Warburg Institute, to give a talk on the recent findings in the Cheops pyramid and the new star cult studies of the Pyramid Texts. It is now hoped that this multi-disciplinary institute will contribute insights into the duality of ancient religions and astrologies.)[22]

Appendix 5

LOGISTICS OF THE SHAFTS IN CHEOPS'S PYRAMID

A Religious Function Expressed with Geometrical Astronomy and Built-in Architecture

by Robert G. Bauval

It is an accepted fact that the design of the Cheops Pyramid – and other pyramids to a lesser degree – incorporates a basic knowledge of geometry and observational astronomy.[1] The intensely geometrical shape of the structure, the precision of design ratios, and its accurate alignments along a precise meridian make this a certainty. Many geometricians who have studied the pyramid agree that harmony of angles and dimension ratios is to be found in the design.[2] Those who have studied its astronomical alignments generally agree that stellar alignments taken at the meridian were the means by which the base of the monument was set out and, as has been shown,[3] the means by which some of the internal features were positioned.[4]

Above all else, however, the monument is intensely religious, with the main cultic purpose of assisting the dead king in his ascent to the sky.[5] In brief, therefore, the monument is a sepulchre with a potent function which, for lack of appropriate terminology, can be said to be astrological.[6] This is a widely accepted consensus and is confirmed by the liturgy of the Pyramid Texts.[7] The religion and rituals of the Pyramid Age were a sky religion, whereby the king became a star and his star soul became established or transferred to the southern stars of Orion and Sirius and to the northern stars, which included the three circumpolar constellations of Ursa Major, Ursa Minor and Draco.[8] The supreme task of the ancient architect was to express these vital elements of the sky religion in the design of the monument. When all is said and done, the pyramid structure was primarily an instrument of rebirth for the departed king.

To achieve this religious function, the architect based his design on simple geometrical principles using right angles and bisected angles fixed with simple mathematical ratios and proportions. This is the common practice in architectural and building engineering design principles, in order to create the ideal functional monument within the constraints of structural considerations and building limitations. Elementary mathematics are bound to be detected by all who study the design of the Cheops pyramid.[9] Yet researchers should not imagine that elementary mathematics was an essential aspect of the pyramid cult; it was merely a tool, albeit probably a sacred tool by which the priestly architect could perform his trade.

The Narrow Shafts

There are four narrow shafts in the Great Pyramid, two from the King's Chamber that emanate northward and southward; two others emanate, also northward and southward, from the Queen's Chamber. These have been discussed in numerous books and articles since 1837.[10] Though they were first thought to be for the purpose of ventilating the internal chambers of the pyramid, the accepted consensus today is that they served a religious purpose of passageways for the ascent of the 'soul' of the dead king.[11] The present writer is a firm supporter of this thesis.[12] There is, however, one main element of the mathematics that needs to be carefully integrated in this thesis if it is to withstand scientific scrutiny: the stellar theory must account for the fact that the architect intended each pair of shafts to emerge at the same horizontal levels on the outside of the pyramid. It is thus important to follow a strategic logic to ascertain, through a series of questions and answers, what the likely intention was of the architect when he opted for this feature.

Mathematical Astronomy or Astronomical Mathematics?

One question that must be answered is: Was the architect briefed to design a monument to express principles of sacred mathematics in the pyramid, or was he briefed to *use* sacred mathematics to provide the pyramid with features that could service the function of the cult, i.e., to assist the departed king to ascend to the sky?

Perhaps the best way to answer this is to use a more modern analogy. In medieval times (and sometimes still today) cathedrals were designed in the shape of a cross generally orientated east. The main entrance was on the west side, at the foot of the cross, which meant that a congregation entering the cathedral would move eastward, thus symbolising the rising of Christ, the east being the place where the celestial orbs rise as the birth star of Christ, 'the star of the east'.[13] Cathedrals were religious monuments intended to service the liturgical aspects of the Christian religion, and the main briefing given to the architect was based on these requirements. The architect developed his design using geometry and mathematics to express in a *symbolic manner* the liturgical function of the cult. The cross was designed in geometrical proportions imbued with deep symbolic meaning: the dome represented the sky vault, the altar was the head of the Christic cross and so on. The architect also used simple observational astronomy to orientate the monument eastward: certain panels towards the sunrise or sunsets and so on.

It stands to reason, therefore, that if a medieval cathedral (such as Chartres in France) is scientifically scrutinised, from its design and orientation will be extracted both sacred mathematics and the elements of simple observational astronomy. But to assume that the main purpose of the architect was to express either is misleading. The correct conclusion would be that the architect used symbolic mathematics and observational astronomy to express the liturgical function of the monument.

The same applies to the Cheops pyramid. A scientific scrutiny will extract the principles of a sacred geometry and certain aspects of observational astronomy, but these are only the tools of the architect's trade and, devoid of religious input, do not elucidate the purpose and function of the monument. A scientific approach is necessary only in that it informs us of the tools and thus of the architectural language through which the religious purpose and function can be understood.

The correct approach to a full understanding of the pyramid design is therefore to make use of elementary mathematics and observational astronomy to extract the symbolic meaning of the design and ultimately link it to the liturgy of the cult. This is also the approach to take in the scrutiny of the shafts in the Cheops pyramid.

A Brief Based on the Religious Function

We know from the Pyramid Texts that both the northern stars and the southern stars were essential aspects of the rebirth rituals and directly related to the celestial destiny of the departed king.[14] It has also been shown by many researchers, Egyptologists and astronomers that the constellations in question were:

a) The northern meridional region: those of Ursa Major,

Ursa Minor and Draco. The last, of course, had its main star, Alpha Draconis, as the pole star of the Pyramid Age (*c.* 2500BC).

b) The southern meridional region: essentially, these were the culmination of the constellations of Orion and Canis Major (which contains Sirius). We must add the constellation of Taurus, including the Hyades, which also had important cultic significance.

All stars, of course, have to be precessed and 'proper motioned' back to the epoch of *c.* 2500BC to meet with the assumed date of the Cheops pyramid.

The religious ritual which took place after the death of the king was essentially a rebirth, as we have said. Some have termed it the Osirian Rites, since ultimately the dead king became an Osiris and departed to the celestial kingdom of this god, in the sky region of Orion.[15] First, however, a variety of ceremonies had to be performed before the dead king was deemed ready to undertake his journey to Orion-Osiris. The most essential was the opening of the mouth during which Horus and his four sons, with ceremonial cutting instruments, opened the mouth of the Osiris-king to induce its rebirth. This ceremony, too, had strong astral connotations but linked to the circumpolar region of the sky. It has generally been accepted that the two ceremonial cutting instruments were shaped to look like the constellations of Ursa Major and Ursa Minor.[16] Another major part of the event was the symbolic birth of a new Horus (the new king), which also had a stellar connotation as 'Horus who is in Sirius-Isis'.[17]

We can therefore safely conclude that the architect's brief was to incorporate in the design of the rebirth chambers architectural elements which would service the essential rituals of the opening of the mouth, the birth of 'Horus who is in Sirius-Isis' and, ultimately, the departure of the soul to the celestial kingdom of Osiris-Orion. In previous articles[18] it was shown that the two southern shafts pointed to Orion's Belt and to Sirius, mythologically Osiris and Isis respectively. The two northern shafts were directed to the pole star, Alpha Draconis, and to the head of Ursa Minor, the celestial adze of Horus, also called the 'adze of Upuaut'.[19] All these alignments work for the same precessed epoch of *c.* 2450BC plus or minus twenty-five years.[20]

Tools and Techniques of the Architectural Design

In considering the techniques of design, we must define the context of the architect. Historically, we are looking at *c.* 2500BC, when the two pyramids at Dashour and that at Meidum were completed by King Sneferu, father of Cheops. The experience acquired in true pyramid design and construction would obviously be related to those pyramids. In accepting that the architect of Cheops used basic geometry to define the scale and proportions and basic observation astronomy to align the base and other features such as the shafts, we must also accept that he had a wider vision based on the past geometrical and astronomical design of the Dashour pyramids and a future vision of the Giza Necropolis as a whole.[21] All these elements had to be linked in one unified architectural vision which, if correct, would be visible in the integrated design and layout of the Dashour and Giza pyramid sites and, ultimately, in the design of the pyramid of Cheops.[22] The final product, the design of the Cheops pyramid had to be linked with the religious purpose of the monument.

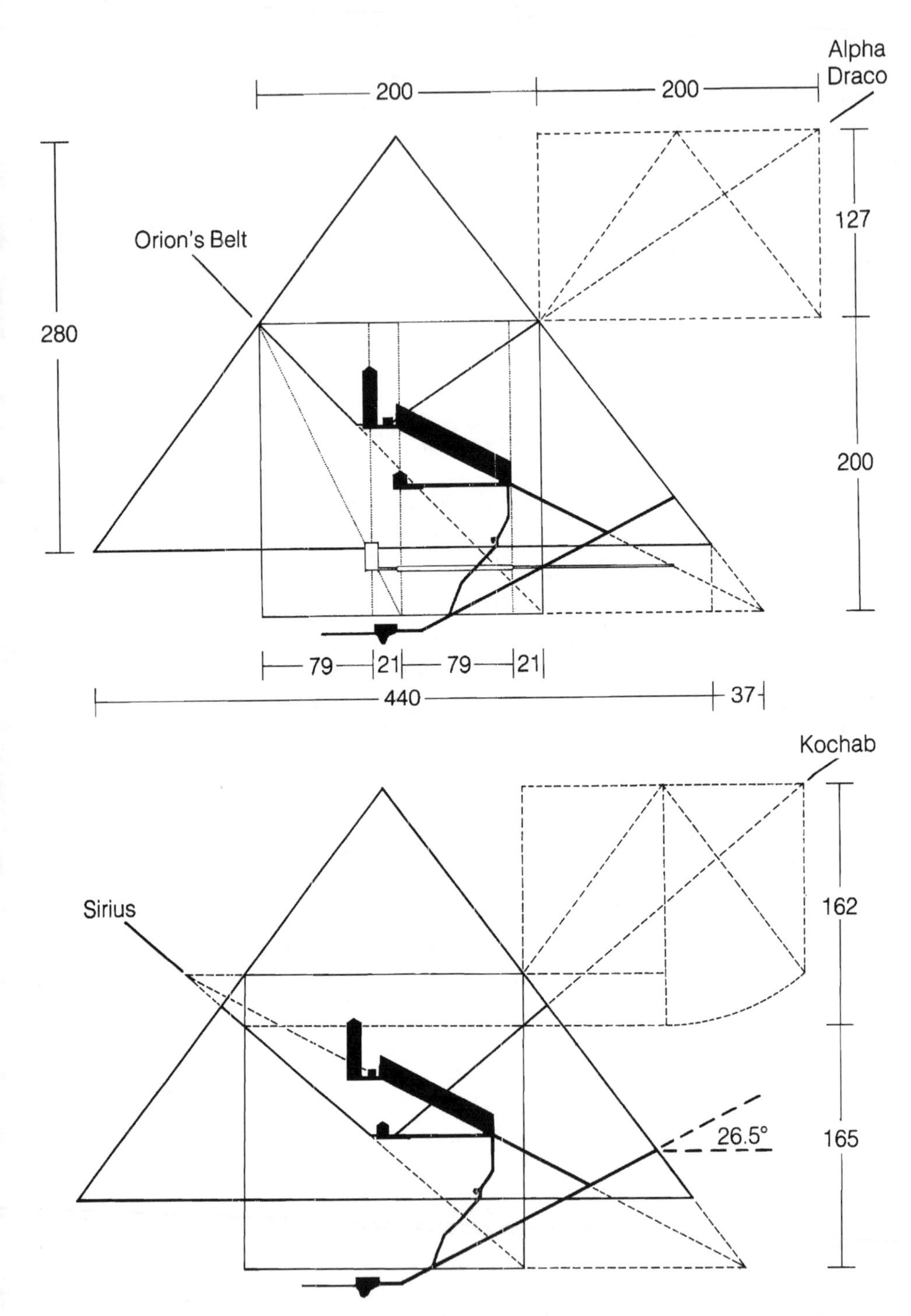

25. ***The astro-geometry of the shafts and chambers inside the Great Pyramid, with the upper culmination position of the stars taken C.2450BC***
All measurements are in royal cubits. 1 royal cubit = 0.5237 metres

Appendix 6

THE HORIZON OF KHUFU

A stellar name for the Pyramid of Cheops

by Robert G. Bauval

In *Discussions in Egyptology* No. 13, it was argued that the three Giza pyramids were constructed to a unified plan, and that the religious motive of the plan was to represent the central region of the sky-Duat, the starry kingdom of Osiris-Orion defined by the three stars of Orion's Belt.[1] Support for this was found in the Pyramid Texts, where the soul of the departed king was said to join Osiris-Orion in the sky,[2] and in the fact that the southern shaft of the King's Chamber was directed to the lower star in Orion's Belt, Al Nitak, at the epoch when the pyramid was constructed.[3]

Link Between the Southern and Northern Shafts

In a recent article,[4] I have shown that the northern shaft of the King's Chamber was directed to the star Alpha Draconis in *c.* 2450BC, and that the northern shaft of the Queen's Chamber was directed to a star in Ursa Minor (Kochab) at its meridian culmination which corresponded to the tip of the celestial 'Adze of Upuaut', which the Pyramid Texts describe as being used by Horus of Letopolis during the ceremony of the opening of the mouth.[5] It was also mentioned that when this specific star in Ursa Minor struck the meridian, so the star Al Nitak (believed to represent Cheops's pyramid) would rise. In the stellar rituals found in the Pyramid Texts we are told that this describes the precise moment of rebirth or rising of the Osiris-king: '. . . Behold, he has come as Orion, behold Osiris has come as Orion . . . O king, the sky conceives you with Orion, the Duat bears you with Orion, you will regularly ascend with Orion from the eastern side of the sky . . .' [PT 820–822].

Furthermore, the actual monument (the pyramid construction) is identified with 'Osiris': '. . . this pyramid of the king is Osiris, this construction of his is Osiris . . .' [PT Utt. 600].

The Name of Cheops's Pyramid

It has been shown by Badawy that the names given to pyramids by the Ancient Egyptians bore strong stellar connotations; Badawy wrote, 'the names of the pyramids of Sneferu, Khufu, Dedefret, Nebre indicate clearly a stellar connotation while those of Sahure, Neferirkare and Neferefre describe the stellar destiny of the ba'.[6] Two such names 'Djedefra is a Sehed star' and 'Nebka is a star' make this certain. Other pyramids have (soul) names; the souls, as many will agree, were thought to be stars.[7] The question, therefore, is whether the name given to Cheops's pyramid could bear a star name, and could this star be identified with Al Nitak, the lower star in Orion's Belt?

There are many variations of the way the name of the Cheops (Khufu) pyramid should be read. The best is given by Edwards as 'Khufu is one belonging to the horizon'.[8] In hieroglyphics, the name appears as

Ȧakhu-t Khufu, the name of the pyramid of Khufu.

[from Wallis-Budge, *An Egyptian Hieroglyphic Dictionary,* vol.I, p.25a; Dover edition 1978].

This means 'The Horizon of Khufu', a name which allows the original hieroglyphic text to speak for itself. We have seen that this pyramid has a likely correlation to Al Nitak, the lower (and larger) star in Orion's Belt; the southern shaft of the King's Chamber was also directed to this specific star when it culminates at the meridian.[9] It also had an adze-shaped shaft,[10] the northern one of the Queen's Chamber, directed to Ursa Minor as it culminates at the meridian when Al Nitak is rising on the horizon. In the Westcar Papyrus, the pyramid is actually called horizon,[11] and in the light of the stellar connotations of such names, it is a 'star in the horizon'. The main stars of the Osirian rebirth were those of Orion, and the evidence is compelling that Al Nitak, poised on the horizon when the cosmic adze strikes the meridian and aligns itself with the northern shaft of the Queen's Chamber, is 'the Horizon of Khufu' (see diagram on p.223).

Appendix 7

THE 'SONS OF RA' AND THE OSIRIAN REBIRTH OF THE PYRAMID KINGS

by R. G. Bauval and R. Cook

1. The 'Osiris' Sons of Ra

It was J. H. Breasted who, in 1912, saw in the Pyramid Texts (*c.* 2300BC) a solar religion which had 'absorbed' an older, and thus quasi-defunct stellar religion during the Pyramid Age.[1] This view, unfortunately, became Egyptological dogma and was adopted by many scholars till this day.[2] In 1966 R. O. Faulkner saw in the Pyramid Texts a strong stellar element but, like Breasted before him, he, too, regarded this as an older and subordinate aspect of the Pyramid Age cult which was, as Breasted had deduced, dominantly solar.[3]

Such a position, however, was challenged in 1964 when A. Badawy and V. Trimble proved that the so-called air-shafts in the King's Chamber of the Great Pyramid were orientated to the stars of Orion's Belt (Osiris) in the south and to the circumpolar stars (Alpha Draconis) in the north.[4] Further evidence of a strong stellar correlation of Orion's Belt and the Giza pyramids came with the studies of R. Bauval in 1989–90, a contributor to this article.[5]

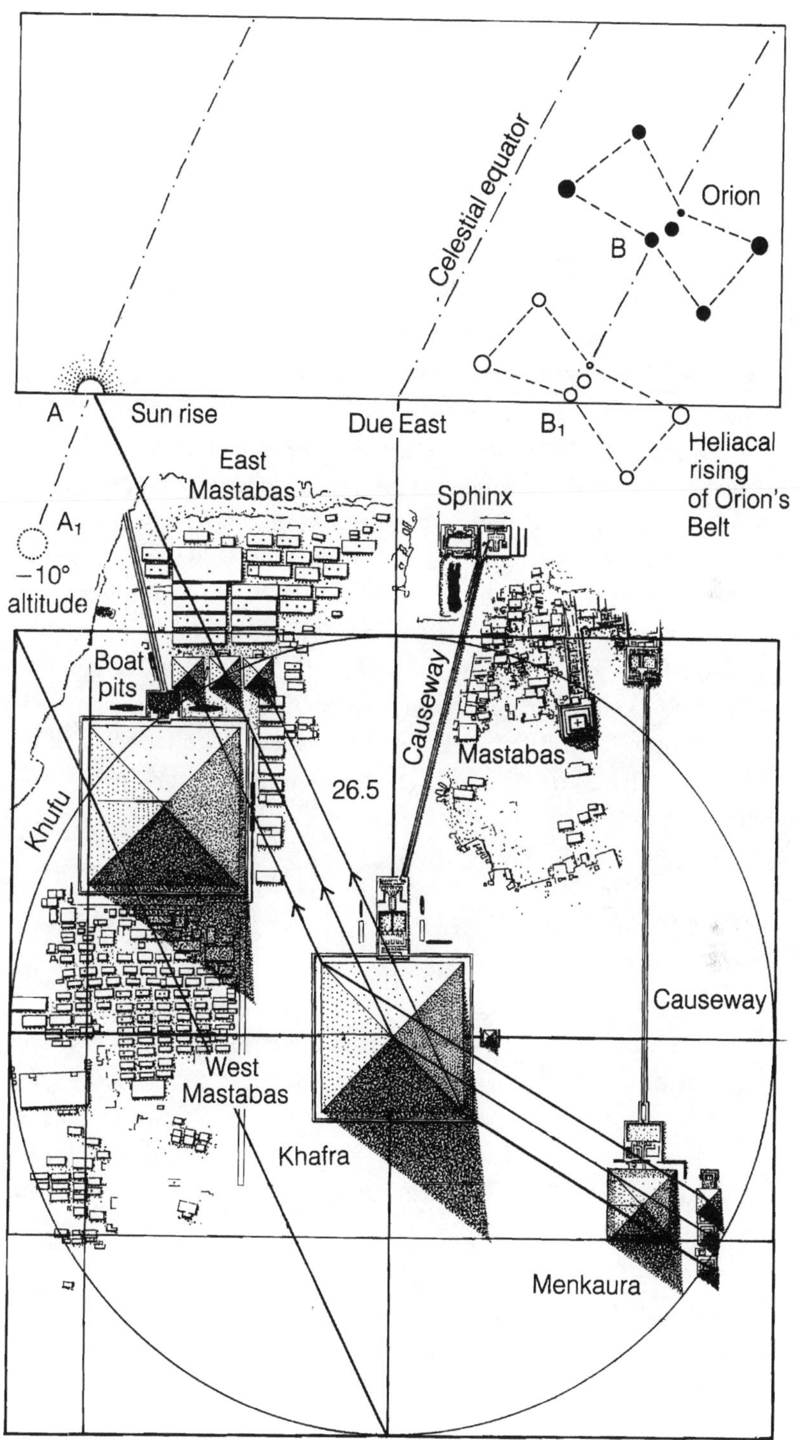

26. The Heliacal Rising of Orion's Belt and the 26.5 degree alignment

Yet seeing the issue as J. H. Breasted did, it always appeared that a 'religious conflict' existed in the Pyramid Age between the state's religious factions of the Pyramid Age, where one faction supposedly favours a 'solar destiny' and the other a 'stellar destiny' for the soul of the departed king. Clearly this is not an acceptable stance to look at the powerful rebirth cult of the Pyramid Age; we do not think that such a religious conflict as imagined by Breasted ever existed. What is more likely the case is that the Pyramid kings saw themselves not as reincarnations of Ra, but rather as the living 'Sons of Ra'; and, as such, they were identified to the divine *progeny* of Ra in the person of 'Horus' when they were alive, and in the person of 'Osiris' when they died. Because Osiris was an *astral god identified to the constellation of Orion*, then the kings expected to undergo an Osirian rebirth which ensured them a stellar destiny with Osiris-Orion and, also, as 'Sons of Ra', in the same way the original 'Osiris' was regarded. This view, which is compatible with the beliefs found in the Pyramid Texts and all other religious texts of other epochs, has the distinct quality of removing the supposed 'conflict' between the solar and stellar ideas of the Pyramid Builders or make us consider a 'solar take-over' of an ancient astral religion during the Pyramid Age.

2. The Heliacal Rising of the 'Horizon of Khufu'

It can easily be shown that the heliacal rising of the Belt of Orion, and more specifically the star Zeta Orionis (Al Nitak), occurred a few weeks before the Summer Solstice in the epoch *c.*2450BC, when the Great Pyramid was built.[6] This meant that the rising point of the sun on this day was at Azimuth 63.5 degrees, that is 26.5 degrees North of East.

3. The 'Cook' alignment of the satellite Pyramids of Giza

The contributor to this article, Robin Cook, an independent researcher on the geometry and layout plan of the Giza Pyramids, has previously shown that the angle 26.5 degrees North of East is the key alignment of the whole Giza complex and especially relates to the three so-called satellite pyramids of Cheops, found on its east side. In short, this alignment directs the whole attention of someone observing the eastern sky to Azimuth 63.5 degrees and also the heliacal rising of Zeta Orionis at the epoch the Great Pyramid was built i.e. *c.* 2450BC. This angle, in view of its cultic links with the 'rebirth' of Osiris-Orion, is unlikely to be coincidental. Furthermore, 26.5 degrees is found also within the Great Pyramid interior design, this being the slope of the descending and also the ascending passages leading to the Chambers of the pyramid. It is well known that the angle of 26.5 degrees is formed by the so-called diagonal of the double-square and was much used by the Ancient Egyptians in the design of monuments. Cook's work has shown, for the first time, that it was also used for the general layout of the Giza Necropolis. This, needless to say, is of enormous interest as it strongly implies a unified master plan for the necropolis as a whole.

4 The Circumpolar Star-Clock

In a previous article by R. Bauval, it was shown how the rising of the star Zeta Orionis in the east coincided with the meridian passage of the star Kochab in Ursa Minor, the target of the northern shaft of the Queen's Chamber.[7] This, it was suggested, explained the name of Cheops's pyramid: 'The Horizon Of Khufu'; furthermore it allocated this pyramid a stellar name, which conforms with the general trend of names of pyramids given by the contemporaries of Cheops such as Djedefra and Nebka.[8]

It follows, therefore, that for the heliacal rising, i.e. rebirth, of Zeta Orionis, the ancient builders could predict this all important event – the 'rebirth' of the star – by observing both the approach of the sun to Azimuth 63.5 degrees (26.5 degrees North of East) and the upper culmination of Kochab. This would strongly suggest that the heliacal rising of stars were not merely determined by waiting impatiently for their rising at dawn – which could be frustrated by haze over the horizon, clouds and excessive refraction – but by cleverly using the circumpolar stars as markers on a sort of 'star clock', with a given meridian upper or lower culmination of specific circumpolar star 'marking', as it were, the time of heliacal rising of another, non-circumpolar star in the east.

NOTES AND REFERENCES

PROLOGUE: The Last Wonder of the Ancient World

1. The Cheops pyramid alone contains about 6.3 million tons of quarried and finely cut rock. The pyramids in the Memphite Necropolis, in the western desert near Cairo, contain over 25 million tons of quarried rock. Stonehenge in England contains about 10,000 tons of roughly hewn rock, thus 2500 times more rock was used to build the Egyptian pyramids and the Great Pyramid is about 600 times more massive than Stonehenge.
2. The stepped structures called ziggurats in Ancient Ur and Babylon may have been begun at the same time (*c.* 2750BC) as the Third Dynasty step-pyramids of Egypt during the Old Kingdom, but the true pyramids (with smooth faces) are an Egyptian invention of *c.* 2550BC. The Mexican pyramids are much younger, dating from no earlier than the first millennium BC. The famous pyramids of the sun and the moon at Teotihuacan in Mexico date from *c.* AD600 (though they may have been built on earlier sites). The best and most recent book on the subject of pyramids around the world is Jean Kerisel's *La Pyramide à travers les Ages* (The Pyramid through the Ages).
3. The Old Kingdom pyramid sites are located over a stretch of desert land some eighty kilometres long and three kilometres wide, very close to modern Cairo, which is known as the Memphite Necropolis.
4. The Arab chronicler, Al Makrizi (fifteenth century AD) in his *Khitat* or Topography (of Cairo) wrote that when Ma'moun found the Great Pyramid contained no treasures, he ordered gold pieces to be put into the sarcophagus in the King's Chamber so that his workers might 'find treasure' and not think their months of strenuous effort in vain. (See Peter Tompkins's *Secrets of the Great Pyramid.* Another entertaining book on the history of pyramid exploration is Leonard Cottrell's *The Mountains of Pharaoh*.)
5. Herodotus, *The Histories*, Book II (paperback, Penguin Books,

Classics series). Many of the facts given by Herodotus on the pyramid are suspect. It was he who, on dubious hearsay 2000 years after the Great Pyramid was buiult, said that Cheops was regarded by the Egyptians as a 'criminal' who treated his people like slaves. Only in the eighteenth century did Europeans begin serious exploration and scientific analysis of the pyramids; their focus was on the Giza pyramids and especially on Cheops's pyramid, in the hope of finding treasure or in an attempt to uncover some religious revelation related to the Bible. In the nineteenth century the British were particularly keen on such theories. After the work of Colonel Howard-Vyse and Perring in 1837, it is generally conceded that Flinders Petrie's *The Pyramids and Temples of Gizeh* (London 1883) was the first serious archaeological work on the Egyptian pyramids. Petrie conducted the first detailed topographical survey and much of his data is still used. Gantenbrink has shown, however, that some of Petrie's inside measurements of the pyramid need fine tuning, especially for the so-called Queen's Chamber shafts. The 'definitive' study on the Egyptian pyramids is Dr I. E. S. Edwards's *The Pyramids of Egypt*.

6. The story of the discovery was seen in many national and international newspapers and periodicals, including the *Daily Telegraph* (7.4.1993), the *Independent* (16.4.1993) *The Times* (17.4.1993), the *Los Angeles Times* (17.4.1993), *Chicago Sun-Times*(23.4. 1993), *Le Monde* (17.4.1993), *Le Figaro* (17.4.1993), *France-Soir* (17.4.1993), the *Daily Mail* (17.4.1993), *Today* (17.4.1993), *Der Spiegel* (19.4.1993), *Stern* (8 July 1993), *Bild, Blick* (16.4.1993), *Bild am Sonntag* (18.4.1993), *Hannoverfsche Allgemeine* (17.4. 1993), *Neue Presse* (17.4.1993), *Hamburger Abendblatt* (17.4. 1993), *Die Welt*, *El Pais*, *Le Matin* (17.4.1993) and several other local papers. The BBC and Channel 4 announced the news on the 16 April 1993, and other TV and radio stations around the world followed.
7. See Bibliography, Bauval R. G.
8. *Independent*, London, 16.4.1993; *Daily Mail* 17.4.1993.
9. *Daily Mail*, London, 17.4.1993; *Today*, London, 17.4.1993.

1 THE GENESIS OF THE ORION MYSTERY

1. Robert K. G. Temple, *The Sirius Mystery*.
2. Near Wad Medani, in the Al Fau region, some 350 kilometres from Khartoum.
3. Sirius has a magnitude of −1.5 and is 8.6 light years away. It rises about one hour after Orion.
4. M. Griaule and G. Dieterlen. 'Un System Soudanais de Sirius', in *Journal de la Société des Africanistes*, XX, fasc. 1, 1950.
5. Alvin Clark first saw it, however, in 1862, through a telescope.
6. Temple, op. cit., p. 1.

7. For a good discussion of the Aten religion, See J. H. Breasted, *Development of Religion and Thought in Ancient Egypt*, pp. 312–343.
8. Porphyry (third century AD) tells us that 'the kings of Egypt . . . had made [Egypt] inaccessible to foreigners' (Porph. De Abstin, IV, 6, Nauck p. 237). This was also said by many ancient historians, such as Diodorus of Sicily (first century AD) (Diod. I 69).
9. The Ancient Egyptians kings did not see themselves as belonging to dynasties, but as a continuous line of divine kings. The notion of separating groups of pharaohs into dynasties is far more recent, and comes from the Egyptian priest and historian, Manetho, who lived in the third century BC when Egypt was under the Ptolemies. Much modern chronology related to Ancient Egypt rests today on Manetho's invention of the dynastic system.
10. Edwards, op. cit., p. 2.
11. In very ancient times the Memphite Necropolis was the land of Sokar or the kingdom of Sokar. Its central region was Rostau, closely identified with the Giza pyramid field. In the Pyramid Age Osiris was personified as Sokar (Edwards, op. cit., p. 10). Common titles for Osiris were Lord of Rostau and Dweller in Rostau. Throughout the pharaonic era, Rostau was considered the main entrance to the afterworld.
12. The city of Annu or On is mentioned in Genesis (41;45) in connection with Joseph and his Egyptian wife, Asenath, daughter of a priest of On. Annu or On (Iwnw in Ancient Egyptian) apparently meant pillar-city (see S. B. Mercer, *The Religion of Ancient Egypt*, p. 127). It was named Heliopolis, probably in the fourth century BC, by the Greeks (Herodotus, op. cit., pp. 2–8.
13. ibid.
14. Aubrey Noakes's *Cleopatra's Needles*, gives a good account of the events that led to the transportation of these obelisks from Egypt to London and New York.
15. Their true hieroglyphic names are best rendered as Khufu (Cheops), Khafra (Chephren) and Menkaura (Mycerinos).
16. This area runs from Abu Ruwash in the north to Dashour in the south. The nearest site to central Cairo is Giza.
17. G. Goyon, *Le Secret des Batisseurs des Grandes Pyramides*, pp. 89–90.
18. See note 12.
19. R. T. Rundle Clark, *Myth and Symbol in Ancient Egypt*, pp. 37–61.
20. ibid., pp. 37–8.
21. ibid., p. 246. Also H. Frankfort, *Kingship and the Gods*, pp. 153, 380 and note 26. R. Bauval, in *Discussions in Egyptology* (henceforth *DE*), vol. 14, p. 7. The idea that the Benben Stone

was probably placed on the On pillar is also expressed by Mercer, op. cit., p. 127.

22. Rundle Clark, op. cit., p. 246.
23. J. Baines, in *Orientalia*, vol. 39, 1970, pp. 389–95. See also Bauval in *DE*, vol. 14, p. 7.
24. Breasted, op. cit. pp. 70–2. Also Edwards, op. cit., p. 282.
25. Edwards, op. cit., p. 284.
26. The hieroglyphic writing was probably invented well before the Pyramid Age. Conservative dating places it around 3000BC. By the time of Cheops it was well developed.
27. The Great Ennead is mentioned in the Pyramid Texts (see Chapter 3); there are allusions to the Great Ennead in earlier tomb writings, although Osiris is not mentioned in the extant material.
28. Rundle Clark, op. cit., p. 246. Clark also sees a link with the planet Venus (a travelling star to the Ancients). See p. 122 for the identification of the soul of Osiris to the stars of Orion. This gives the Benben a rather stellar character (the matter is fully discussed in Chapter 11).
29. See E. A. Wallis-Budge, *Osiris and the Egyptian Resurrection*, vol. 1, for a full discussion.
30. The term *mastaba* was coined by Auguste Mariette in the 1860s; it reminded him of the bed-like sitting areas, called Mastabas, seen outside the homes of rural Egyptians.
31. 'Mansions of Eternity' are, of course, the step-pyramids and, eventually, the great true pyramids.
32. Edwards, op. cit., p. 19.
33. ibid., p. 19.
34. ibid., pp. 34–70.
35. ibid., p. 34 and p. 284. Edwards says, 'Imhotep's title "Chief Of The Observers" . . . became the regular title of the High Priest of On.'
36. W. Lethaby, *Architecture, Mysticism and Myth*, p. 129.
37. ibid.

2 THE MOUNTAINS OF THE STAR GODS

1. Edwards, op. cit., pp. 292–3.
2. True pyramids certainly did have a Benben at the top. (Edwards, p. 282). There is every reason to believe that the same applied to the earlier step-pyramids.
3. There are several pyramidions. The best example, in the main hall of the Cairo Museum, is the pyramidion of Amenemhet III.
4. Helio-Polis, literally sun city in Greek.
5. A. Moret, *Le Nil et la Civilisation Egyptienne*, 1926, p. 203; Edwards, op. cit., p. 282.
6. Edwards, ibid.

7. The thirty million tons estimate is based on data provided in *Atlas of Ancient Egypt* by J. Baines and J. Malek, p. 140. It does not include the temples, causeways, ramps etc., which formed part of the permanent and temporary construction operations. The density of limestone was taken as 2400 kg/M3.
8. Baines and Malek, op. cit., pp. 135 and 140. See also Malek, *In the Shadow of the Pyramids*, inside cover.
9. Edwards, op. cit., p. 2.
10. For a detailed study on the Meidum pyramid, see K. Mendelssohn, *The Riddle of the Pyramids*. See also Edwards, op. cit., pp. 71–3.
11. Mendelssohn, op. cit., p. 40, gives 850,000 tons for Zoser. Edwards, op. cit., p. 92, gives 9 million tons for the two Dashour pyramids, but also includes the casing for the Meidum pyramid.
12. The southern pyramid or 'bent' pyramid of Dashour is actually rhomboidal in shape, the bottom half of a pyramid with a slope of 54 degrees, on which is the top half of a pyramid with a slope of 43.5 degrees. The northern pyramid at Dashour has a slope of 43.5 degrees.
13. The southern pyramid has retained most of its casing stones. Seen from afar, the northern pyramid also looks semi-intact. In contrast, the much younger pyramid of the Twelfth Dynasty, that of Amenemhet III, looks like a heap of rubble.
14. Malek, op. cit., p. 47.
15. A. Badawy, *A History of Egyptian Architecture*, vol. 1, p. 124.
16. Edwards, op. cit., p. 73.
17. ibid., p. 78.
18. ibid.
19. ibid.
20. ibid., p. 92. Since the first edition of his book in 1947, Edwards has revised his views on Sneferu in the light of new data and discoveries. His 1993 edition allocates the two Dashour pyramids to Sneferu.
21. ibid.
22. ibid.
23. ibid., p. 93.
24. ibid.
25. See Chapter 3. As for the Great Pyramid's lack of official inscriptions, some graffiti were found inside the 'relief chambers', in which some have read the name Khufu. This is the strongest 'evidence' in the arsenal of the Egyptologists to show that it belonged to Khufu (Cheops). For a different view, see W. R. Fix, *Pyramid Odyssey*, pp. 75–89.
26. The Great Pyramid remained the world's tallest structure (146 metres) until 1888, when the Eiffel Tower was planned in Paris (300 metres). But such comparisons are unrealistic; the mass of the Eiffel Tower is 7175 tons compared with 6.3 million tons for the

Great Pyramid. The Empire State Building is 381 metres high, and is estimated to contain not more than 300,000 tons of permanent structural material.

27. Edwards, op. cit., p. 152.
28. Wallis-Budge, *The Mummy*, p. 10.
29. See Appendix 2.
30. Edwards, op. cit., p. 289; Malek, op. cit., p. 124; J. B. Sellers, *The Death of Gods in Ancient Egypt*, p. 338.
31. Baines and Malek, op. cit., p. 36. Here the start of Khufu's reign is given as *c.* 2551BC, and comes closest to my own estimates (see *DE*, vol. 26, 1993).
32. Letter from I. E. S. Edwards to R. G. Bauval dated 27.1.1993.
33. Bauval in *DE*, vol. 26, 1993, p.5.
34. Edwards, op. cit., p. 284. Edwards also says that the high priest wore a coat decorated with stars.
35. ibid. Contrary to the current consensus, we see ancient Heliopolis (Annu) as being a 'wisdom school', where the focus was on the observation and recording of the motion of stars.
36. It is not impossible that Imhotep was still alive when Zoser's pyramid was completed and Sneferu came to power. The dramatic change in design suggests either that this was the case, or that Imhotep was succeeded by someone he trained.
37. We now use our new computed date given in *DE*, vol. 26, p. 5.
38. Mark Lehner, 'The Development of the Giza Necropolis: the Khufu Project', in *MDAIK*, 1985, band 41, Tafeln 1–3. Also in Newsletters of *JARCE*, nos. 131, 135. Also see Lehner 'Some observations on the Layout of the Khufu and Khafra Pyramids', in *JARCE* XX.
39. Lehner, *MDAIK*, op. cit., pp. 114–18.
40. The Dashour pyramids are also visible from Giza on a bright day from the high level of the Giza plateau. From the tarmac canal road which passes near Dashour, the pyramids appear suddenly when you are almost next to them on the east.
41. P. D. Ouspensky, *A New Model of the Universe*, pp. 350–1.
42. Edwards, op. cit., p. 98.
43. This estimate is widely accepted by Egyptologists. Dr Jean Kerisel (see note 2 in Prologue) thinks this is too high in view of the main 'voids' caused by the wide jointing in the core of the monument. His estimate is closer to 4.7 million tons (Kerisel, op. cit., p. 67). Both values are, however, largely theoretical since there is no telling how many joint voids there are. The estimate of 6.3 million tons allows for filling joints with debris.
44. Ouspensky, op. cit., p. 352.
45. Edwards, op. cit., pp. 291–2.
46. ibid., p. 104. See also Chapter 12 of this book.
47. Waynman Dixon was an employee of the North England Iron Company. He was a good friend of Petrie's father, who corres-

ponded with him and met him in 1873. In 1865 Piazzi Smyth also met Dixon and became good friends. In 1875 Dixon and his brother, John, were recruited by Sir Erasmus Wilson (founder of the Egyptian Exploration Fund in 1883 and its first president) to organise the transport of the obelisk, Cleopatra's Needle, now in London. See Epilogue.

48. These shafts were closed until Waynman Dixon discovered them in 1872. Dixon and Dr Grant, his colleague, lit fires at their mouths in the Queen's Chamber to see if they penetrated to the outside of the pyramid, but no smoke was seen. Mysteriously, the smoke in the southern shaft seemed to disappear into the monument itself (C. Piazzi Smyth, *The Great Pyramid; its secrets and mysteries revealed*, p. 428).
49. The southern shaft is about sixty-five metres long, but there is no telling how much space lies beyond it. The northern one, though still not fully explored, is at least 24 metres long.
50. Ouspensky, op. cit., p. 354.
51. Plato's *Timaeus and Critias* is a dialogue which deals essentially with the so-called Atlantis myth.
52. Proctor was a well-respected astronomer. He founded the popular science magazine, *Knowledge*.
53. Some 1500 tons of granite was imported from Aswan (1000 kilometres to the south) to construct the King's Chamber and its five relief chambers on top. Why limestone was not used, as happened in the Queen's Chamber, remains a mystery. The King's Chamber is about forty metres up the monument, so it must have been very difficult to construct without lifting devices and machines.
54. This is confirmed by Dr Richard Parkinson of the British Museum. See also A. Gardiner's article 'The Secret Chambers of Thoth' in *JEA*, 11, 1925, pp. 2–5. Also E. Hornung in *ZAS*, 100, 1973, p. 33.
55. Tompkins, op. cit., pp. 218, 284.
56. For a recently published discussion on the quarries of Zawyat Al Aryan and Abu Ruwash, see Edwards, op. cit., 1993 edition.
57. The figure is likely to be closer to 26 million tons if we allow for temporary ramps and filling material.
58. Malek, op. cit., p. 117.
59. Sahura's pyramid (Fifth Dynasty), although also in pitiful condition, is probably the best preserved.
60. Smaller pyramids were built until well into Christian times (in Meroe in the Sudan). In Memphis, construction went on at Lisht, Hawara and Dashour until the Thirteenth Dynasty, but these are poor structures compared with the Fourth Dynasty prototypes.
61. Badawy, op. cit., p. 143.
62. Malek, op. cit., p. 119.
63. Edwards, op. cit., p. 152.

64. J. A. Kane, *The Ancient Building Science*, introduction.
65. M. Isler, in *JARCE*, XXVI, 1989. Isler is a solar theorist though Edwards, Lauer and others have long shown that the Great Pyramid was aligned with star sightings.
66. Edwards, op. cit., p. 247.
67. ibid., pp. 248–51.
68. R. O. Faulkner, 'The king and the star-religion in the Pyramid Texts', in *JNES*, XXV, p. 153.
69. ibid.
70. In 1974 Mark Lehner gathered the so-called Edgar Cayce readings into a book, *The Egyptian Heritage*. Cayce was a mystic who advocated in his readings, delivered in trance, that the Great Pyramid's construction was started in 10450BC by survivors of Atlantis. On the back cover of the book, Lehner seemed to agree with this view. For more on Edgar Cayce, see T. Sogrue, *There is a River*. (ARE Press, Virginia 1963.)
71. M. Lehner, in *MDAIK*, band 41, p. 109.
72. R. J. Cook, in *The Giza Pyramids*, a design study.
73. J. A. R. Legon, in *Report of the Archaeological Society of Staten Island*, vol. 1, NY 1979.
74. Legon, in *DE*, vol. 10, 1988, pp. 33–9.
75. ibid., p. 38.
76. ibid.
77. Apparently he will do so in future. He is, however, following a purely geometrical and mathematical course of reasoning to discuss the esoteric meaning of the layout plan.
78. Cook, op. cit.
79. ibid.
80. Robin Cook, who also prepared the diagrams for *The Orion Mystery*, believes that the motives were entirely religious and expressed through a combination of geometrical and astronomical data; we agree. The ancient designers were obviously like architects today – trained in many disciplines, including geometry, astronomy, religion, history, symbology and so on, and whose art was to know how to express the whole in architectural design to provide the monument with religious function and countenance.

3 THE DISCOVERY OF THE PYRAMID TEXTS

1. Tewfik-Pasha was an Anophile and was apparently initiated into freemasonry, and once served as the Grand Master of the Egyptian Grand National Lodge (Paul Naudon, *Histoire générale de la Franc-Maconnerie*, (Office du Livre, 1987 ed., p. 224). In 1881 he asked the British government to come to his aid and depose Arabi Pasha, his minister of war, who was plotting a coup against him. A British force, led by Wolesley, arrived in Alexandria in July 1882 and shelled the city. Wolesley's forces closed with Arabi's army at

Tell Al Kebir and Arabi was defeated. Technically, Egypt became a British protectorate from then on.

2. Several incidents against Europeans were reported. In June and July 1882 Arabi's supporters went on a rampage in Alexandria, killing Europeans and looting their villas and shops. This apparently justified Wolesley's shelling of the city from his warships anchored off the ancient harbour.
3. G. Maspero, in *Rec. Trav.*, vol. v, Fasc. I–II, p. 157.
4. P. Montet, *Isis: ou à la Recherche de l'Egypte Ensevelie*. Montet gives an excellent account of Mariette's life and work. Mariette seemed to get himself in rather tight spots with public figures; he brawled with Verdi over *Aida*, and was in Empress Eugenie's bad books for refusing her a gift of Egyptian jewellery. He also got into many academic brawls, one violent between him and the German Egyptologists under Lepsius.
5. Montet, op. cit., p. 48. Mariette received a fund of FF30,000 to resume work at Saqqara. At the time, this was a small fortune.
6. See note 25, Chapter 2.
7. Montet, op. cit., pp. 81–2.
8. Maspero, op. cit., p. 157.
9. ibid.
10. ibid.
11. ibid. Also, Montet op. cit., wrote that Mariette, with his great experience, should have known better. We can only assume that bias blinded his judgement.
12. Maspero, in *Bull. Eg. Serv.* II, vol. 6.
13. ibid.
14. Maspero, in *Rec. Trav.*, III, p. 179.
15. ibid.
16. Breasted, op. cit., p. 102.
17. W. R. Dawson and E. P. S. Uphill, *Who Was Who In Egyptology*, p. 38.
18. Breasted, op. cit., introduction, p. vii.
19. Dawson and Uphill, op. cit., p. 38.
20. Breasted, op. cit., p. 93.
21. ibid.
22. Breasted, op. cit., pp. xiv–xv.
23. ibid., pp. 312–43.
24. ibid.
25. ibid., p. xi.
26. E. A. Wallis-Budge, *The Egyptian Book Of The Dead*, p. ix.
27. ibid., p. xii.
28. Edwards, op. cit., p. 177.
29. The three major works mentioned are: K. Sethe, *Die Altägypttischen Pyramidentexte*, 3 vols., 1908–12; S. B. Mercer, *The Pyramid Texts in Translation & Commentaries*, 4 vols., 1952; A. Piankoff, *The Pyramid of Unas*, Bollingen Series XL. 5, 1968.

30. R. O. Faulkner, *The Ancient Egyptian Pyramid Texts*. OUP ed., 1969.
31. Faulkner, op. cit., p. v.
32. Among them, Frankfort in 1948, and J. B. Sellers in 1992 (see Sellers, op. cit., pp. 7–9).
33. Letter from C. Keller to Robert Bauval dated 8.10.1986.
34. Rundle Clark, op. cit., p. 13.
35. ibid., p. 12.
36. ibid., p. 13.
37. Letter from a Swiss-German Egyptologist to Robert Bauval dated 9.12.1986.
38. Letter from C. Keller, 8.10.1986.
39. S. Hassan, *Excavations At Giza*, vol. VI, part i, p. 43.
40. Mercer, *The Pyramid Texts*.
41. Faulkner, op. cit., p. vii.
42. Mercer, *The Religion of Ancient Egypt*, pp. 25 and 112.
43. ibid., pp. 121–2.
44. Faulkner, op. cit., p. vii.
45. 'R.O. Faulkner, The king and the star-religion . . .', in *JNES*, XXV, 1966, pp. 153–61.
46. Letter from I. E. S. Edwards to Robert Bauval dated 20.7.1986.
47. As note 42.
48. H. Frankfort, *Ancient Egyptian Religion*, 1961 ed., preface.
49. A. Piankoff, *The Tomb of Ramesses VI*, Bollingen Series XL. 1.
50. Sellers, op. cit. Sellers is a UCLA graduate and studied Egyptology at The Oriental Institute in Chicago.
51. ibid.
52. ibid., pp. 173–5.
53. Frankfort, *Ancient Egyptian Religion*.
54. Sellers, op. cit., p. 8.

4 LET THE PYRAMID TEXTS 'SPEAK'

1. Between 305BC and AD642 it was one of the great centres of learning. Among the numerous scholars who lived and studied in Alexandria then were the mathematicians Euclid and Heron; the astronomers Eratosthenes, Hipparchus, Timochares, Posidonius, and Ptolemy; the philosophers Theophrastus and Clement of Alexandria, and the theologian Arius.
2. The word 'copt' comes from the Greek *Aigyptos* which means Egypt (the Greeks gave Egypt its modern name; the Arabs call their country Misr). Copts or Aigyptii were the natives who became Christians during the Graeco-Roman epoch. The Coptic church is still strong in Egypt and has its own pope or patriarch. In Egypt in March 1993, I briefly met the Coptic bishop of Cairo, his holiness Bishop Musa (Moses).
3. This is part of the so-called Osirian Mysteries of Ancient Egypt.

The belief in an Osirian afterlife was first a privilege of kings; it was gradually democratised until everyone was entitled to an Osirian rebirth. The rituals were complex and, in the case of kings, may have taken several months after the death of the monarch. The mummy (a modern word coming from the Arab noun *mummia* which means pitch or tar) was regarded as the Osirianised version of the dead person.

4. Mercer, *The Religion of Ancient Egypt*, p. 25.
5. ibid., p. 112.
6. Sellers, op. cit., p. 70.
7. Hassan, op. cit., pp. 276–317. Hassan gives a full description of the Duat.
8. O. Neugebauer and R. Parker, *Egyptian Astronomical Texts*, vol. 1, pp. 24–5.
9. ibid.
10. ibid.
11. Rundle Clark, op. cit., p. 122.
12. Mercer, *The Religion of Ancient Egypt*, p. 270.
13. E. A. Wallis-Budge, *Osiris and the Egyptian Resurrection*, vol. 1, p. 107.
14. Obtained by Skyglobe 3.5 astronomical program. This date is verified by others.
15. During the epoch of the Fourth Dynasty, the heliacal rising of Sirius occurred some five to seven days after the summer solstice. At this time the sun rises near azimuth 63.5 degrees and Sirius, at the epoch, rose near azimuth 116.5 degrees.
16. A well-known fact in Egyptology. The religious calendar thus began its New Year's Day when Sirius was rising heliacally. Because of the extra quarter-day difference between the true year and the calendrical year of 365 days, this meant that the religious calendar lost a day every four years. Both the religious calendar and the civic calendar would synchronise again, of course, every 1461 years ($4 \times 365.25 = 1461$). This period of 1461 is often called the Sothic Cycle.
17. Wallis-Budge, *Osiris*, vol. 1, gives the various sources for the Osirian myth. Plutarch's *De Iside et Osiride* is the most detailed from classical times (*c.* AD50).
18. Maat was represented by a winged goddess wearing a feather on her head.
19. Frankfort, *Kingship and the Gods*, Part I.
20. Faulkner, 'The king and the star-religion . . .', pp. 153–61.
21. By Skyglobe 3.5 and allowing for proper motion in declination −1.21 arcseconds per year. J. Legon (see Bibliography), using his own precessional program, gets −21 degrees 38.38 minutes. He gets −20.85 degrees for epoch 2500BC; this value is also given by astronomer and navigator H. R. Mills in his book, *Positional Astronomy and Astro-navigation made Easy*, p. 232. Mills gives

−20.83 degrees for 2500BC, which is the same as −20 degrees 49 minutes. For epoch 2450BC (the date of Khufu's pyramid), Skyglobe 3.5, with the application of proper motion to its reading, gives −20 degrees 30 minutes with an estimated value of plus or minus five minutes for hand-reading precision with the mouse on the screen. Very much the same result is obtained with EZ Cosmos program when adjusted for proper motion in declination.

22. J. Greaves, *Pyramidographia*, 1646, p. 73.
23. Abbé le Mercier, *Description de l'Egypte, composée sur les memoires de M. de Maillet*, Paris 1735.
24. Jomard, *Description de l'Egypte*. Edition Panckoucke, 1821–1829, tome IX, p. 491.
25. J. S. Perring, *The Pyramids of Gizeh*. Part I: The Great Pyramid.
26. W. F. Flinders Petrie, *The Pyramids and Temples of Gizeh*, (1990 ed. by Histories & Mysteries of Man Ltd.), p. 29.
27. Piazzi Smyth, *The Great Pyramid*, p. 428.
28. Flinders Petrie, op. cit., p. 24.
29. *Archeologia*, vol. 293, September 1993, p. 6. See also *Stern* magazine, Number 28, July 1993, pp. 24–5.
30. J. Capart, *Études et Histoires*, I, Bruxelles 1924, p. 182.
31. G. Steindorff, *Egypt*, Baedeker 1929, p. 140.
32. Edwards, op. cit., (1961 ed.), p. 126.
33. J. Vandier, *Manuel d'Archaeologie Egyptienne*, Tome II, Paris 1954, p. 88.
34. A. Badawy, 'The Stellar Destiny of Pharaoh and the so-called Air-shafts in Cheops's Pyramid', in *MIOAWB*, band 10, 1964, pp. 189–206.
35. In *MIOAWB* band 10, 1964, (Trimble) pp. 183–7 and (Badawy) pp. 189–206.
36. Badawy, op. cit., p. 190.
37. ibid.
38. ibid.
39. ibid.
40. V. Trimble, 'Astronomical Investigation concerning the so-called Air-Shafts of Cheops's Pyramid', *MIOAWB*, band 10, 1964, pp. 183–7.
41. ibid., p. 187.
42. I. E. S. Edwards, 'The Air-channels of Chephren's Pyramid', in *Studies in Honor of Dows Dunham*, p. 55–7.

5 THE GIZA PLAN

1. The pyramidion was found in 1902 by Maspero near the pyramid of Amenemhet III at Dashour (*Ann. Serv.* III, 1902, p. 206). It is made of finely polished black granite and is remarkably well preserved. It weighs about four tons.
2. A setting-out engineer is responsible for fixing the grid system

from which the builder will develop the construction on the site. A theodolite (20-arcseconds reading), ranging rods and poles, 30 or 100 metres steel tape, tilting or 'dumpy' level, plumb bobs (8 oz.), set of set-squares, nylon line (hank), straight edge (steel), plus a variety of materials such as claw hammer, builder's square, stakes etc., are the typical tools of the trade. Omitting the theodolite, tilting or 'dumpy' level and items made of steel, the Ancient Egyptian setting-out engineer would have required all the others for the accuracy he achieved. Levelling course would have been achieved by slope ratios and temporary water channels; straight lines on the ground, through stellar sighting at the meridian (Edwards, op. cit., pp. 250–1). Much debate among scholars plagues this issue.

3. It contains some 250,000 blocks. Assuming a 20 blocks per day, this would take 34 years. To bring it down to 10 years, we have assumed 69 blocks per day, about 7 blocks per hour, far too high, in my opinion, for an epoch without wheeled transport and without lifting machines.
4. Preliminaries are the various non-permanent works on a building project, such as workers' accommodation, temporary access roads, temporary stock piles, offices, drainage, workshops water supply and so forth. On large engineering projects, especially in a remote area (such as the western desert near ancient Memphis), these can easily amount to 15–20 per cent of the full works.
5. J. P. Lauer, *Observations sur les Pyramides*, p. 99.
6. ibid.
7. ibid, pp. 99–124.
8. Z. Zäba, *L'Orientation Astronomique* . . .
9. It is easier, of course, for a setting-out engineer to fix one grid line instead of three, and during building operations each pyramid would serve as a sighting for the next.
10. J. Phaure, *Introduction à la Geographie Sacrée de Paris*, 3eme edition, Borrengo, p. 29.
11. *National Geographic*, vol. 180, No. 2, August 1991, pp. 122–34.

6 GIZA AND THE BELT OF ORION

1. J. Lacouture, *Champollion: Une Vie de Lumières*, ed. Grasset and Fasquelle, Livre de Poche 1988, pp. 428, 456. Apparently shouted by Champollion on 14 September 1822 to his elder brother, when he realised that he had worked out how to decipher the Egyptian hieroglyphs.
2. E. A. Wallis-Budge, *Heaven and Hell*, preface p. x and pp. 131–5, 348. See also Hassan, op. cit., p. 315.
3. Edwards, op. cit., p. 10.
4. Carved on the so-called Shabaka Stone, British Museum item 498.
5. Shabaka Texts, line 18c.

6. Hassan, op. cit., p. 302.
7. Sellers, op. cit., p. 164.
8. ibid., p. 165.
9. Rundle Clark, op. cit., p. 27.
10. Tompkins, op. cit., pp. 218, 284.
11. Baines and Malek, op. cit.
12. Rundle Clark, op. cit., p. 97.
13. Goyon, op. cit., p. 198.
14. Rundle Clark, op. cit., p. 108.
15. M. Lichtheim, *Ancient Egyptian Literature*, vol. 1, p. 204.
16. R. O. Faulkner, *The Book of the Dead*, (glossary) Rostau.
17. ibid., Spell 173, p. 172.
18. R. O. Faulkner, *The Book of the Dead*.
19. Hyginus Poet. Astr. 2.32; A. B. Cook, *Zeus* vol. 2, p. 481.
20. Diodorus, I, 12,5.
21. Eusebius, Praep. Ev. III, 3,6.
22. R. H. Allen, *Star Names: Their Lore and Meaning*, p. 216.
23. Wallis-Budge, *The Egyptian Book of the Dead*, p. cxxiii.
24. One pyramid of the Fifth Dynasty (Unas); four pyramids of the Sixth Dynasty; and in three small pyramids of queens.
25. Edwards, op. cit., (1991 ed.), pp. 288–9.

7 THE STAR CORRELATION THEORY

1. Dr Edwards retired in 1974, though he is still very active. He lives with his wife not far from Oxford.
2. Edwards, 'The Air-channels . . .' pp. 55–7.
3. Letter dated 8.1.1985. I have since met Dr Malek on several occasions. He is at present directing the task of recording all Egyptological works (archives) for the Griffith Institute, and agreed that the contents of his letter could be disclosed.
4. The present apparent motion of Orion's Belt, when plotted on the declination changes, decreases in the negative range (i.e., upward towards the celestial equator). From *c.* AD2500, the motion will be downwards.
5. This is a good pocket mini-computer, it costs about $US100.
6. J. A. R. Legon, 'A Ground Plan at Giza', in *DE* 10, 1988. I understand that Legon intends to provide a religious interpretation of the plan, but not concerned with Orion-Osiris.
7. R. G. Bauval, 'The Seeding of the Star Gods', in *DE* 16, 1990.
8. ibid.
9. See note 6 of Prologue.

8 THE BROTHER OF OSIRIS

1. R. G. Bauval, 'A Master Plan . . .' in *DE* 13, 1989.
2. Edwards, op. cit., (1991 ed.), pp. 288–9.

3. One is extremely damaged and it appears to most visitors, at first, that only three pyramids are at Abusir.
4. Edwards, op. cit., p. 152.
5. Malek, op. cit., inside cover.
6. Approximate distance. Estimate from *Atlas of Ancient Egypt* and Egyptian government survey map.
7. Having the sites two kilometres from each other is, in engineering logistics, a rather curious choice for two monuments being built at the same time for the same king. In many temporary works it would have meant doubling the resources, which could have been avoided if the two pyramids were sited, say, 500 metres or so from each other such as at Giza.
8. R. G. Bauval in *DE* 26, 1993, p. 5.
9. Edwards, op. cit., p. 93. This is a theory put forth by Dr R. Stadelmann. Most Egyptologists assume a reign by Sneferu of twenty-four years.
10. Sellers, op. cit.
11. ibid., p. 11.
12. ibid., p. 174.
13. ibid.
14. Catalogue 2000.0 was used for this book.
15. E. C. Krupp, *In Search of Ancient Astronomies*, pp. 186–190.
16. J. Cornell, *The First Stargazers*, p. 92.
17. Sellers, op. cit., p. 116.
18. Lichtheim, op. cit., p. 51.
19. A thesis was written in 1986 and a copy provided to Dr Edwards, who discussed it with me.
20. Lichtheim, op. cit., p. 51.
21. ibid.
22. Sellers, op. cit., p. 90.
23. In the Pyramid Texts Geb is often said to be the earthly father of Osiris. He is the legitimate consort of Nut, the sky goddess and mother of Osiris. In the Pyramid Texts, however, Nut seems to have had explicit sexual encounters with Ra (or Atum-Ra), the sun god.
24. Lehner, *The Egyptian Heritage*, pp. 128–9.
25. ibid.
26. In the northern hemisphere, from the equator to the north pole. The farther from the pole, the higher is the celestial equator.
27. Sellers, op. cit., p. 116.
28. Edwards, op. cit., p. 152.
29. ibid.
30. P.B. 3033 in the East Berlin Museum. Most of the Westcar Papyrus is kept in vaults. I was allowed to see and photograph it by Dr Wilddung and his assistant, Dr Muller, in September 1993.
31. Edwards, op. cit., p. 153.

32. ibid.
33. Plutarch, *The Life of Alexander the Great*. The story is told by many chroniclers of Alexander's life. In one of the versions (Plutarch), it is said that the last native king of Egypt, Nectanebo II, changed into a snake (Zeus-Ammon) and made love to Olympia. Zeus-Ammon was much venerated by the Greeks in the time of Alexander, and his oracle at Siwa (western Egypt) was symbolically linked to the oracle of Zeus at Dodona by Herodotus (*The Histories*, II).
34. M. Lyttleton and W. Forman, *The Romans: Their Gods and Their Beliefs*, Orbis 1984, p. 29, shows how Julius Caesar claimed Venus (Aphrodite) as the ancestress of the Julian house. Other legends make his real mother become pregnant by a snake, probably symbolising, like Alexander before him, Zeus-Ammon in his snake form.
35. Louis XIII was believed impotent by many.
36. This phrase was coined by the Hermetic Philosopher, Tomasso Campanella, in 1636. He apparently drew the solar horoscope for the future sun king of France, Louis XIV.
37. Edwards, op. cit., p. 152.
38. This is evident from the so-called Coffin Texts, the Middle Kingdom's version of the older Pyramid Texts.
39. This idea of the fourteen pieces of Osiris's body and the pyramids of Memphis is not new. It was mentioned on several occasions by researchers (see e.g. T. Holland, *Freemasonry from the Pyramids of Ancient Times*, p. 14, where he states '. . . that Osiris having fourteen tombs for various parts of his dismembered body, fourteen pyramids must have been devoted to them').
40. A. W. Shorter, *The Egyptian Gods*, p. 39. Also Wallis-Budge, *The Egyptian Book of the Dead*, p. li.
41. Wallis-Budge, 'Osiris . . .', vol. 1, p. 99.
42. Shorter, op. cit., p. 43.
43. See Denderah Zodiac for example. (Neugebauer and Parker *Egyptian Astronomical Texts*, vol. *Plates*.)
44. This is evident from the position of Orion (Mithra). See statue of 'Mithra killing the Bull' in the British Museum, (upper galleries, Ancient Roman Section).
45. Typically in such depictions, Orion the Giant or Orion the Hunter holds the head of the bull with his left hand and is about to club it with the mace, held in his right hand. The head of the celestial bull is the Hyades cluster of stars.
46. Star 311 was on the left (north of Aldebaran) at rising point looking east.
47. Shorter, op. cit., pp. 41–3.
48. ibid.
49. ibid.
50. Sellers, op. cit., p. 116.

9 INTERMEZZO AT THE PYRAMIDS

1. A. G. Gilbert (ed.), *The Hermetica*, Solos Press 1992, foreword, pp. 5–29.
2. A. G. Gilbert, *The Cosmic Wisdom beyond Astrology*.
3. See Chapter 10.
4. I. E. S. Edwards, 'Do the Pyramid Texts suggest an explanation for the Abandonment of the Subterranean Chamber of the Great Pyramid?' for J. Leclant's *Festschrift* in 1994.
5. Also with us was Frau Marion Krause-Jach, a pharmacist from Berlin, a very good friend of mine who was visiting Egypt. Marion was most helpful during conferences given by Rudolf Gantenbrink in London (22 April 1993) and Paris (21 June 1993).
6. See Chapter 4, note 40.
7. *DE* 26 came out in May 1993. *DE* 27 is due in late September 1993.
8. See Chapters 11 and 12.
9. Channel 4 *News at Seven*, 16.4.1993.
10. See Prologue.
11. Prof. Kerisel was the first person to propose a scheme for ventilating the Great Pyramid (see 'Chauffage Ventilation Conditionement', in *Revue de l'Association des Ingénieurs Climatique*, Ventilation et Froid, No. 12, Decembre 1990, p. 54).
12. Rundle Clark, op. cit., p. 263.

10 THE GREAT STAR-CLOCK OF THE EPOCHS

1. Papyrus British Museum 10371/10435; translation by R. Parkinson, *Voices of Ancient Egypt*, p. 65.
2. Rundle Clark, op. cit., p. 263.
3. ibid.
4. See Chapter 3.
5. Wallis-Budge, *The Egyptian Book Of The Dead*, p. xii, fn. 1.
6. The length of the southern shaft (the 'Gantenbrink' shaft) in the Queen's Chamber makes this obvious (see Prologue).
7. Krupp, op. cit., p. 186. Yet to be fair, Krupp does treat Ancient Egyptian astronomy with respect and felt that 'we may yet be surprised by what we find among the tombs and temples of the Nile' in an astronomical sense. It was, surprisingly, Richard Parker and Otto Neugebauer who did not see any 'scientific' application in the astronomy of the Ancient Egyptians. Astronomer James Cornell, however, felt that all science in ancient Egypt was subjugated by religion.
8. S. Mayassis, *Mystères et Initiations de l'Egypte Ancienne*, Chap. I, pp. 1–13.
9. Diodorus, I, 69.

10. Strabo, *Porphyre*.
11. Herodotus, *Histories*, II, 2–8.
12. Dion Chrystomenos, XI, 37s.
13. Schwaller de Lubicz, *Sacred Science*, p. 11.
14. Letter from Dr G. Haeny (Swiss Archaeological Institute, Cairo) to Robert Bauval dated 9.12.1986.
15. Schwaller de Lubicz, op. cit., p. 16.
16. Mendelssohn, op. cit., quoted by Cornell, op. cit., p. 84.
17. A. Weigall, *A History of the Pharaohs*, vol. I, pp. 1–15.
18. ibid. See also R. Parkinson, *Voices of Ancient Egypt*, p. 48. Parkinson shows a graffito from Wadi Hammamat which depicts cartouches of kings of the Fourth Dynasty, including Khufu, Redjedef, Khafra, Hordjedef, Baufre and names of certain goldsmiths.
19. Schwaller de Lubicz, op. cit., p. 87. Also see Gilbert, *Hermetica*, p. 20.
20. Diodorus, I, 44.
21. Schwaller de Lubicz, op. cit., p. 87.
22. C. Cerf and Y. Navasky, *The Experts Speak: the Definitive Compendium of Authoritative Misinformation*, Pantheon Books, NY 1984, pp. 9–10. This is a must for all who face academic and 'expert' opposition.
23. ibid., pp. 3–4.
24. Rundle Clark, op. cit., p. 246.
25. Abatte-Pasha, 'Le Phenix Egyptien', in *Bull. Inst. Eg.*, II, 4, p. 11.
26. Rundle Clark, *The Legend of the Phoenix*, Part I, pp. 1–17.
27. G. W. Oosterhout, 'The Heliacal Rising of Sirius', in *DE* 24, 1992. Also M. F. Ingham, 'The Length of the Sothic Cycle', in *JEA*, 55, pp. 36–40 for a detailed discussion on the Sothic Cycle and period of invisibility of Sirius.
28. Sellers, op. cit., p. 204.
29. ibid., p. 193.
30. Tompkins, op. cit., pp. 174–5.
31. Sellers, op. cit., p. 174. G. Santillana has also strongly argued the case (see note 30).
32. Schwaller de Lubicz, op. cit., pp. 175–8.
33. We are now, in 1993, at almost highest declination – a few centuries (minutes for the precessional clock) from the midnight hour of the great Orion-Osiris cycle.
34. Strabo, *Porphyre*.
35. Herodotus, *Histories*, II, 2–8.
36. Plato, *Timaeus and Critias*, 22–26. See also edited version by Betty Radice, Penguin Books 1977, appendix, pp. 146–67.
37. Timaeus, 41E–42D.
38. W. Scott (trans.), *Hermetica*, Solos Press ed., p. 228.
39. Walter Scott first advocated this in his edition of the *Hermetica* (1924). We understand that recently Warburg Institute

scholars were read a paper by Peter Kingsley, a research student, 'Hermetica: The Egyptians' Origins & Background' (26.5.1993); Kingsley, it seems, believes that the true source of the *Hermetica* is Egyptian.

40. They were writing in Greek but were very likely Greek-educated Egyptians. Walter Scott believed that the author of Asclepius I (one of the books of the Hermetic Writings) was 'probably an Egyptian by race' (*Hermetica*, p. 228).
41. From at least the time of Herodotus (fifth century BC) the Greeks believed that the priests of Heliopolis had acquired a sacred wisdom which was recorded in holy books, books which they ascribed to the god Thoth, the inventor of hieroglyphic sacred writings and geometry. It was this god, Thoth, whom the Greeks later called Hermes Trismegistos and attributed the Hermetic Writings to him. Thoth was a supposed messenger or secretary of God, and was thus imbued with infinite wisdom; his writings, therefore, would reveal true gnosis (divine knowledge). In Egyptian drawings, Thoth is a human figure with the head of an ibis, the symbol of wisdom. He is often seen holding a writing board and a marker, in his role of dispatcher of divine messages and recorder of human deeds, especially those which would be weighed in judgement in the afterlife court of Osiris (the so-called Great Hall of Judgement). In this last capacity, Thoth would establish whether the deceased had acquired gnosis and was entitled to a place in the cosmic Kingdom of the Dead ruled by Osiris.
42. *Hermetica*, Asclepius III, Solos Press ed., p. 136.
43. ibid.
44. Fix, op. cit., p. 99. See also Lehner, *The Egyptian Heritage*, pp. 131–2.
45. T. Sugrue, *There is a River*, ARE Press, 1988 ed., p. 356. Cayce died on 3 January 1945.
46. These belong, we understand, to the Association for Research and Enlightment (ARE), with headquarters at Virginia Beach, 67th Street and Atlantic Avenue, VA 23451. They have a large library and operate centres around the world.
47. Sugrue, op. cit., p. 468. Also J. E. Furst, *Cayce's Story of Jesus*, Berkeley Books, NY 1976 ed., p. 101.

11 THE SEED OF THE PHOENIX

1. Rundle Clark, *Myth and Symbol* . . . op. cit. p. 246.
2. Herodotus, *Histories* II, 73.
3. A. Lucas, *Ancient Egyptian Materials and Industries*, p. 299.
4. Baines, in *Orientalia*, vol. 39, 1970, pp. 389–95.
5. 'Ben' or 'bin', literally 'son', is still commonly used in the Arab world and in Israel to denote 'son of . . .'

6. Edwards, op. cit., p. 282.
7. Rundle Clark, *Legend of the Phoenix*, pp. 5–6.
8. Rundle Clark, *Myth and Symbol in Ancient Egypt*, p. 246.
9. Rundle Clark, *Legend of the Phoenix*, p. 17.
10. ibid.
11. ibid., p. 15; Edwards, op. cit., p. 282.
12. Edwards, op. cit., p. 282.
13. This is produced by the shock waves in the atmosphere.
14. For a full study on meteoritic cults, see various articles by G. A. Wainwright in *JEA*, vol. 18, pp. 3–15; vol. 17, pp. 185–95; vol. 21, pp. 152–70. See also C. Daremberg and E. Saglio, *Dictionnaire des Antiquités Grecs et Romaines*, 'Elagalabus' and 'Baetylia'; Cook, *Zeus*, III, part I, pp. 881–903.
15. I am grateful to Vagn Buchwald, an authority on iron meteorites at the Instituttet For Metallaere, Lyngby, Denmark. He supplied me with wonderful photographs of meteorite 'Morito' which he had taken. I am also indebted to Brian Mason of the Smithsonian Institution in Washington DC for permitting a reproduction of meteorite 'Willamette' taken in the 1940s. 'Willamette' is now at the American Museum of Natural History in New York.
16. See note 14.
17. ibid.
18. ibid.
19. Pausanias, *Desc. Greece*, III, 22,1; L. R. Farnell, *The Cults of the Greek States*, vol. I, chapter v.
20. Pliny, *Nat. Hist.*, II, 59.
21. P. K. Hitti, *History of Syria*, London 1951, p. 312.
22. See note 14.
23. ibid.
24. See note 14. There is also mention of a sacred meteorite in the Bible concerning the symbol of Diana of Ephesus 'which fell from heaven' (Acts 19:35).
25. G. A. Wainwright, in *Ann. Serv.*, xxviii, 177.
26. G. A. Wainwright, 'Some Aspects of Amun' in *JEA*, p. 147.
27. G. A. Wainwright, 'Iron in Egypt' in *JEA*, 18, 1933, pp. 3–15.
28. ibid. The word *bja* also means meteoritic iron, since this was the only form of iron available in *c.* 2400BC.
29. G. A. Wainwright, 'Iron in Egypt', p. 14. This is mentioned by Plutarch (AD50), in *De Iside et Osiride*, sect. 2, Teubner's ed., Moralia, ii, p. 536.
30. Since the 1920s the idea has occurred to many. Wallis-Budge seems to have been the first to make the connection (see note 31). As far as I can make out, I am the first to suggest that it was an 'oriented' i.e. conical, iron meteorite of about fifteen tons (see my article in *DE* 14, 1989, pp. 5–17).
31. E. A. Wallis-Budge, *Cleopatra's Needles*, London 1926, Chapter

1; J. Leclant and J. P. Lauer (eds.), *Le Temps des Pyramides*, pp. 79 and 336.

32. Goyon, op. cit., p. 225. He estimates a weight for the pyramidion of Cheops's pyramid of 15.89 tons. Kerisel, on the other hand, estimates 4–6 tons, but wonders if it ever existed (*La Pyramide à Travers les Ages*, p. 83).
33. A large fallen meteorite, contrary to popular belief, is not burning hot when it strikes the ground but *cold*. This is because of sub-zero temperatures in outer space and only the outer skin is heated by friction when it is flying through the earth's atmosphere. In a state of rest, the outer layer cools quickly.
34. Frankfort, *Kingship and the Gods*, pp. 153, 380 and note 26.
35. See Chapter 4, note 11.
36. Wallis-Budge, *The Mummy*, p. 175.
37. These are probably titles and deal with the same person at different stages of his career, i.e., Horus as child, as heir and as king. On the role of Horus as the real son of the king see Edwards, op. cit., p. 32.
38. These were called Hapy, Imseti, Duamutef and Kebhsenuf. In the Pyramid Texts their main function was to help in the opening of the mouth ceremony and lifting up the dead king to the sky. This is a depiction very suited in astral terms, as Wainwright has shown, to the low culmination of the four stars of the 'head' of Ursa Minor, as they slowly 'scoop' over the horizon in the north and then ascend to an altitude some fifteen degrees over the pole in *c.* 2450BC.
39. G. A. Wainwright, 'A Pair of Constellations' in *Studies for F. Ll. Griffith*. Wallis-Budge, *The Egyptian Book of the Dead*, p. cxxiv. As gods of the four cardinal points they symbolised the four pillars that hold the canopy of the sky. They also represented the four corners of a chamber serving a funerary ritual (Sellers, op. cit., p. 248). In the funeral rites, the four sons of Horus are also symbolised by the so-called canopic jars which held the lungs (Hapy), liver (Imseti), stomach (Duamutef) and intestines (Kebhsenuf) of the deceased person.
40. Pyramid Texts, lines 1983–4.
41. Wainwright, 'Iron in Egypt'.
42. ibid., p. 11.
43. B. Scheel, *Egyptian Metalworking and Tools*, p. 17.
44. Wainwright, 'Iron in Egypt', p. 11.
45. Lauer, op. cit., p. 106, fn. 1. (*The Pyramid Texts in Translation and Commentaries*, Commentaries section).
46. Egyptologist and astronomer from Prague, Zbynek Zäba came up with almost the same conclusion in 1953. His work, *L'Orientation Astronomique dans l'Ancienne Egypte et la Precession de l'Axe du Monde*', has been widely accalimed in the Egyptological world. Lauer, in his review of Zäba's work (see note 49), stresses Zäba's

revelation that a similar instrument to that used by the ancient pyramid builders to sight the northern polar stars was used, symbolically, for the ceremony of the opening of the mouth (see Lauer, p. 108; Zäba, p. 72). Apparently the instrument was placed on a wooden cube and, linked to a plumb line, allowed a very 'precise alignment to the stars'. Interestingly, three instruments were found in these shafts in 1972 which suggested something of the kind (see Epilogue). Wainwright ('A Pair of Constellations') correctly pointed out that two 'adzes' were used in the ceremony, one by Horus and the other by his four sons.

47. I am indebted to Gantenbrink for a plan view of the two northern shafts which show in detail the circuit or 'kinks' of these shafts.
48. See Chapter 12.
49. Goyon, op. cit., p. 160.
50. ibid., p. 89; Wainwright, 'Iron in Egypt', p. 6.
51. Goyon, op. cit., pp. 88–93.
52. ibid.
53. ibid., pp. 160–2.
54. Wainwright, 'Iron in Egypt', pp. 6–15.
55. Baines and Malek, op. cit.
56. Sellers, op. cit., p. 164.

12 THE ROADS OF OSIRIS

1. Sesostris I (Twelfth Dynasty) was the hero-king of the Middle Kingdom. He was the son of Amenemhet I, who was apparently assassinated at court. Amenemhet I and Sesostris I built their pyramids at El Lisht, a few kilometres south of Dashour.
2. Letter from Dr G. Haeny (Swiss Archaeological Institute, Cairo) to Robert Bauval dated 9.12.1986.
3. Breasted, op. cit., p. 71; Parkinson, op. cit., p. 41. The word 'Benben' is also translated as 'Pyramidion'.
4. Breasted, p. 71.
5. ibid.
6. ibid., p. 203.
7. Leiden Museum Papyrus No. 344. See also Breasted, op. cit., p. 204.
8. Leiden Museum Papyrus No. 344 (henceforward LMP 344); 6–12.
9. LMP 344; 5, 10.
10. Breasted, op. cit., p. 204.
11. LMP 344; 11, 6–8.
12. In view of their common antiquity (pre-Pyramid Age) and common religious function, the alignment is unlikely to be a coincidence.
13. Sellers, op. cit., also gives a translated quote, p. 164.
14. Wainwright, 'Iron in Egypt', pp. 6–11.
15. Goyon, op. cit., pp. 89–90.
16. Edwards, op. cit., pp. 250–1 and Lauer, op. cit., pp. 99–124,

only consider the north stars. But setting-out engineers, as is well known, 'toss' or reverse their sighting instruments (theodolites) 180 degrees to verify the true alignment. This is sometimes called a backsight, performed as a check. Thus a similar sighting point to the south would secure the precision achieved by the ancient builders.

17. Farouk I abdicated in 1952. Goyon settled in France and became head of research of the Centre National des Recherches Scientifiques (CNRS). He had been a student of Pierre Montet, the famous French Egyptologist of the College de France.
18. Goyon, op. cit., p. 93.
19. Strabo, 19–xvii, I, 30.
20. Herodotus, *Histories*, II, 15–17.
21. Goyon, op. cit., p. 92.
22. ibid.
23. Lichtheim, op. cit., p. 204.
24. Goyon thinks they might have been slightly concave discs.
25. Mercer, *Religion of Ancient Egypt*, p. 121.
26. Bauval, in *DE* 14, 1989, p. 7.
27. Louvre: Stela item C 30.
28. The djed or tat column, probably kept in Heliopolis, was sacred to Osiris. It might have represented his backbone (see Wallis-Budge *Osiris* . . ., vol. 1, pp. 51–3).
29. Lichtheim, op. cit., p. 203.
30. The best Osirianised example is, of course, the solid gold coffin of Tutankhamen in the Cairo Museum.
31. An article, 'The Horizon of Khufu' by Robert Bauval, is due for publication in *DE* in 1994.
32. It seems unlikely that a monument which took several decades to build and taxed the nation to the limit, was to be used only once for a rebirth ritual. See opinion given in Tompkins, op. cit., p. 236 and pp. 256–8 for other ritualistic functions associated with rebirth.
33. Bauval *DE*, 26, 1993, p. 5; *DE*, 27, 1993.
34. I am indebted to Rudolf Gantenbrink for letting me see this 'niche' on the video film taken by UPUAUT I inside the southern shaft of the King's Chamber.
35. This was pointed out to me by Dr Haeny (see note 2); he was apparently informed by a 'friend'. A quick glance at a map shows it to be quite correct.
36. Probably as early as the outset of the Fourth Dynasty. Certainly, by the time of Sesostris I the temples at Heliopolis had suffered pillage and damage, because Sesostris ordered a full reconstruction programme.
37. Kerisel, op. cit., p. 83. He reminds us that Diodorus (*c.* first century BC) reported that the top of the pyramid appeared like a 'platform'; but Diodorus's account describes the pyramid nearly 2400 years after Cheops.

38. Westcar Papyrus, Berlin 3033.
39. Lethaby, op. cit., Architectural Press ed. 1974, pp. 38–9, 'according to Brugsch, the sun temple at Heliopolis had a sacred sealed chamber in the form of a "pyramid", called, Ben-ben . . .'
40. Sugrue, op. cit.
41. ibid., p. 393

EPILOGUE

1. All quotes are from these personal letters, notes and diaries. The originals are kept in the archives of the Royal Observatory at Edinburgh by Mr Angus Macdonald, the Librarian.
2. Colonel Howard Vyse. *Operation Carried On At The Pyramids Of Gizeh in 1837*. James Frazer, London 1840, pp. 275–277.
3. Lucas A. *Ancient Egyptian Materials And Industries*. Histories & Mysteries Of Man Ltd London 1989, p. 237.
4. El Sayed El Gayar and M. P. Jones 'Metallurgical Investigation of an Iron plate found in 1837 in the Great Pyramid at Gizeh, Egypt' in Journal of the Historical Metallurgy Society, vol. 23, 1989, pp. 75–83.

APPENDIX 1

1 W. Flinders Petrie, *The Pyramids and Temples of Gizeh* 1883, p.53.
2. W. S. Smith, *The Art and Architecture of Ancient Egypt*, Penguin Books 1958, p. 30.
3. Alexander Badawy, *A History of Egyptian Architecture*, Giza 1954, p. 163.
4. Although α Cephei and α Lyrae (Vega) will come to within 4 degrees of the pole in AD7500 and 14000, they will not be nearly as accurate pole stars as Polaris and α Draconis, whose closest approaches to the pole are only about 30 minutes away.
5. Robert H. Baker, *Astronomy*, New York 1950, p. 57.
6. Otto Neugebauer and Richard A. Parker, *Egyptian Astronomical Texts*, I. The Early Decans, London 1960, p. 25.
7. ibid., p. 110.
8. S. R. K. Glanville, *The Legacy of Egypt*, Oxford 1942, pl. 32.
9. Paul V. Neugebauer, *Tafeln zur astronomischen Chronologie*, I. Sterntafeln. Leipzig 1912, pp. 8, 20.
10. ibid. pp. 21–82.
11. This culmination was, of course, rendered invisible by daylight during about half the year. It would have been visible about 2700BC from late July to early January.

APPENDIX 2

1. The software PC programs that are easy to use and quite accurate for down to epoch 4000BC are Skyglobe 3.5; EZ Cosmos; Starmap 2.10 High Precision. These are relatively well priced. Many others are available at different prices.
2. A. Hirshfeld and R. W. Sinnot, *Sky Catalogue 2000.0*, vol. 1. Cambridge University Press 1982. Introduction p. xiv.
3. ibid., pp. 119, 121 and 124.
4. Dr J. O'Byrne, of the Chatterton Astronomy Department of the University of Sydney, says about the precessional calculations for Epsilon Orionis (the central star of Orion's Belt) that 'the precessed position for Epsilon Orionis has been calculated assuming a very small value for proper motion in both co-ordinates (rather than zero) to see the effect. The result is a difference of 4 seconds of RA and 65 seconds of declination, which is probably unrealistically large.
5. I. E. S. Edwards, *The Pyramids of Egypt*. Penguin, 1993 ed., pp. 247–51.
6. J. B. Sellers, *The Death of Gods in Ancient Egypt*. Penguin, 1992 ed., p. 194.
7. J. E. Manchip-White, *Ancient Egypt*. George Allen & Unwin Ltd., 1970, p. 138.
8. O. Neugebauer and R. Parker, *Egyptian Astronomical Texts*. Brown University, 1964, vol. 1.

APPENDIX 3

1. So, for example, Erman, *Die Literatur der Aegypter*, 70, 72.
2. This determinative may indicate a house, a room, or any object, like a box, which contains in the way that a house contains.
3. The photograph is indistinct; see Möller, *Hieratische Lesestücke*, I, 6.
4. *JEA*, III, Pl. XII, between 96–7.
5. On 54 of his Commentary.

APPENDIX 4

1. E. A. Wallis-Budge, *An Egyptian Hieroglyphic Dictionary*, II, p. 942.
2. ibid., p. 946.
3. Edwards, op. cit., 1991 ed., p. 1.
4. Faulkner, *The Book of the Dead*, p. 12.
5. ibid.
6. ibid.
7. Edwards, op. cit., 1991 ed., p. 13.
8. ibid., p. 14.

9. Rundle Clark, *Myth and Symbol* . . ., p. 102.
10. Coffin Texts, ii, p. 104.
11. Neugebauer and Parker, op. cit., vol. I.
12. ibid., p. 41.
13. ibid.
14. Carlsberg, I, Part I, A.I, 1–6, the Nut picture.
15. ibid., G.IV, 26–9.
16. Dramatic Texts, VI, pp. 3–6.
17. ibid., pp. 38–43.
18. Neugebauer and Parker, op. cit., vol. I, p. 73.
19. Bauval, *DE*, 14, p. 12; *DE*, 16, pp. 21–8; Wernes Honig, *DE*, 14, p. 52.
20. Wallis-Budge, *The Mummy*, p. 175.
21. Frankfort, *Kingship and the Gods*.
22. Santillana and von Deschend, op. cit., for further reading on ancient sky religions and duality.

APPENDIX 5

1. For astronomy, see Bauval in *DE*, 13, 14, 16, 26 and 27; geometry and mathematics, see J. A. R. Legon in *DE*, 10 and 14. See also Edwards, op. cit., pp. 245–51.
2. Robin Cook, op. cit., Seven Islands ed., 1992.
3. Edwards, op. cit.
4. Such as the shafts, ibid., p. 285.
5. See Bauval, *DE*, 13 and 14.
6. *Archeologia*, No. 283, September 1993, p. 6.
7. Bauval, in *DE*, 13, 14, 16.
8. Bauval, in *DE*, 26.
9. Legon, in *DE*, 10 and 14; R. Cook, op. cit; R. Gantenbrink (unpublished).
10. Edwards, op. cit., p. 285; Bauval, in *DE*, 13, 16, 26 and 27.
11. A. Badawy 'The Stellar Destiny of Pharaoh . . .' in *MIOAWB*, Band 10, 1964, pp. 189–206.
12. Bauval, in *DE*, 13, 16, 26 and 27.
13. Jean Phaure, *Geographie Sacrée de Paris*, introduction; Lethaby, op. cit.
14. Faulkner, 'The King and the Star-Religion . . .', pp. 153–61.
15. Pyramid Texts, e.g. lines 820, 882, 2180.
16. *DE*, 28 and 29.
17. Pyramid Texts, line 632.
18. Bauval, in *DE*, 13 and 16.
19. Pyramid Texts, line 13.
20. Bauval, in *DE*, 26 and 27.
21. Legon, in *DE*, 10 and 14; Legon, 'The Geometry of the Bent Pyramid', in *Göttinger Miszellen*, No. 116, 1990, pp. 65–73.
22. Legon, as note 21; see also R. Cook, op. cit.

APPENDIX 6

1. *DE*, 13, pp. 7–18.
2. Faulkner, 'The King and the Star-religion . . .', in *JNES*, xxv, 1966, pp. 153–61.
3. *DE*, 26, pp. 5–6.
4. ibid.
5. *DE*, 28.
6. Badawy, 'The Periodic System . . .', in *JEA*, 63, p. 58.
7. Bauval, in *DE*, 14, pp. 5–16.
8. Edwards, op. cit., p. 295.
9. *DE*, 26, pp. 5–6.
10. *DE*, 28.
11. Goyon, op. cit., p. 41.

APPENDIX 7

1. J. H. Breasted, *Development Of Religion and Thought In Ancient Egypt*. Un. Penn. Press 1972 ed., pp. 101–2.
2. I. E. S. Edwards, *The Pyramids Of Egypt*. Penguin 1993 ed., p. 282.
3. R. O. Faulkner, 'The King and the Star Religion in the Pyramid Texts', in *JNES* vol. 25, 1966, pp. 153–161.
4. A. Badawy, 'The Stellar Destiny of Pharaoh and the so-called Air-shafts of Cheops' Pyramid', in *MIDAWB* band X, 1964, pp. 189–206. Also V. Trimble, 'Astronomical Investigations concerning the so-called Air-shafts of Cheops' Pyramid', in ibid., pp. 183–7.
5. R. G. Bauval, various articles in *DE* vols. 13, 14, 16, 26, 27 and three others forthcoming in early in 1994.
6. Computer programs such as Skyglobe 3.6, Starmap or EZ Cosmos can demonstrate this visually.
7. R. G. Bauval, 'The Adze of UPUAUT' and also 'The Horizon of Khufu' in *DE* journal (early 1994 ref. not available yet).
8. A. Badawy. 'The Periodic System of Building a pyramid' in *JEA* 63, 1977, p. 58.

BIBLIOGRAPHY

ABBREVIATIONS

Ann.Serv.	Annales du Service des Antiquités de l'Egypte
Arch.Ast.JHA	Archaeo astronomy, Journal of the History of Archaeology
Bull.Eg.Serv.	Bulletin Egyptologie Service
Bull.Inst.D'Eg.	Bulletin de l'Institut d'Egypte
Bull.Soc.Fr.D'eg.	Bulletin de la Société Française d'Egyptologie
Bull.Inst.Fr.	Bulletin de l'Institut Français d'Archéologie Orientale
DE	Discussions in Egyptology
JARCE	Journal of the American Research Centre in Egypt
JEA	Journal of Egyptian Archaeology
JNES	Journal of Near Eastern Studies
MDAIK	Mitteilungen des Deutschen Archeologischen Instituts, Abteilung Kairo
MIOAWB	Mitteilungen des Instituts für Orientforschung Akademie der Wissenschaften zu Berlin
Rec.Trav.	Recueil de Travaux Relatifs à la Philologie et l'Archéologie Egyptiennes et Assyriennes
Rev.Arch.	Revue Archéologie
Rev.D'Eg.	Revue d'Egyptologie
ZAS	Zeitschrift für AgyptischeSprache und Altertumskunde

ABATTE-PASCHA. 'Le Phenix Egyptien', in *Bull.Inst.D'Eg.*, II, 4, pp. 9–15

ABBOTT, P. *Geometry*. Hodder & Stoughton, London, 1977 ed.

ALLEN, J.P. 'The Pyramid Texts of Queens Jpwt and Wdbtn (j)', in *JARCE*, xxiii, 1986, pp. 1–25

ALLEN, R.H. *Star Names: Their Lore and Meaning*. Dover Publications

Inc., New York, 1963 ed.
ANTONIADI, E.M. *L'Astronomie Egyptienne depuis les Temps les plus Reculés*. Imprimerie National, Paris 1934
BADAWY, A. 'The Stellar Destiny of Pharaoh and the so-called Air-shafts in Cheops's Pyramid', in *MIOAWB*, band 10, 1964, pp. 189–206
— 'The Periodic System of Building a Pyramid', in *JEA*, 63, 1977, pp. 52–8
BAINES, J. & MALEK, J. *Atlas of Ancient Egypt*. Nathan, Paris, 1981 (French ed.)
BAUVAL, R.G. 'A Master Plan for the Three Pyramids of Giza based on the Configuration of the Three Stars of the Belt of Orion' in *DE*, 13, 1989, pp. 7–18
— 'Investigations on the Origins of the Benben Stone: Was It an Iron Meteorite?' in *DE*, 14, 1989, pp. 5–17
— 'The Seeding of the Star Gods: A Fertility Ritual inside Cheops's Pyramid?' in *DE*, 16, 1990, pp. 21–9
— 'Cheops's Pyramid: A New Dating using the Latest Astronomical Data', in *DE*, 26, 1993, pp. 5–7
— 'The Upuaut Project: New Findings in the Southern Shaft in the Queen's Chamber of Cheops's Pyramid' in *DE*, 27, 1993
— 'The Adze of Upuaut: The Opening of the Mouth Ceremony and the Northern Shafts in Cheops's Pyramid' (projected 1994), with A. Gilbert
— 'The Horizon of Khufu: A Stellar Name for Cheops's Pyramid' (projected 1994)
— 'Logistics of the Shafts in Cheops's Pyramid: A Religious Function expressed with Geometrical Astronomy and Built-in Architecture' (projected 1994)
BLACKER, C. & LOEWE, M. *Ancient Cosmologies*. George Allen & Unwin Ltd., London 1975
BRANDON, S.G. F. *Religion in Ancient History*. George Allen and Unwin Ltd. London 1973
BREASTED, J.H. *Ancient Records of Egypt*. Histories and Mysteries of Man Ltd. London, 1988 ed.
— *Development of Religion and Thought in Ancient Egypt*. University Of Pennsylvania Press, Philadelphia, 1972 ed.
BRECHER, K. & FEIRTAG, M. *Astronomy of the Ancients*. MIT Press, Mass. 1979 ed.
BRIGHTY, S.G. *Setting-Out: A Guide for Engineers*. Crosby Lockwood Staples, London 1975
BUCHWALD, V.F. *Handbook on Iron Meteorites*. UCLA Press, Berkeley 1975
BURNHAM, Jr. R. *Burnham's Celestial Handbook*. Dover Publications Inc., New York 1978
CAUVILLE, S. *La Théologie d'Osiris à Edfou*. Institut Français d'Archéologie Orientale du Caire, 1983
CHAMDOR, A. *The Book of the Dead*. Garrett Publications, 1966

CHURTON, T. *The Gnostics*. Weidenfeld & Nicolson, 1987
COOK, A.B. *Zeus*, vols. 1, 2 and 3. Cambridge 1940
COOK, R.J. *The Pyramids of Giza*. Seven Islands, Glastonbury, 1992
CORNELL, J. *The First Stargazers, An Introduction to the Origins of Astronomy*. The Athlone Press, London 1981
COTTRELL, L. *The Mountains of Pharaoh*. Robert Hale Ltd. London 1956
CRICHLOW, K. *Order in Space; a Design Source Book*. Thames & Hudson, London 1973 ed.
DARESSY, M.G. 'Une Ancienne Liste des Décans Egyptiens', in *Ann.Serv. Tom.*, i, pp. 79–90
DAVID, R. *Mysteries of the Mummies; the story of the Manchester University Investigation*. Book Club Associates, London 1978
DAVIDSON, M. *The Stars and the Mind*. Watt & Co. London 1947
DAVIS, V.L. 'Identifying Ancient Egyptian Constellations', in *Arch.Ast.*, vol. 9, JHA xvi, 1985, S102
DAWSON, W.R. & UPHILL, E. P. *Who Was Who in Egyptology*. Egyptian Exploration Society, London 1972
DORMION, G. & GOIDON, J.P. *Kheops: Nouvelle Enquête*. Editions Recherche sur les Civilisations, Paris 1986
DUNHAM, D. 'Building an Egyptian Pyramid', in *Archaeology*, vol. 9, 1956
DUNHAM, D. & SIMPSON, W.K. *The Mastaba of Queen Mersyankh III, G 7530–7540*. Museum of Fine Arts, Boston 1974
EDWARDS, I.E.S. *The Pyramids of Egypt*. Penguin Books, London 1993
— 'The Air-Channels of Chephren's Pyramid', in *Studies in Honor Of Dows Dunham*, Boston 1981, pp. 55–7
ERMAN, A.A. *Handbook on Egyptian Religion*. Archibald Constable & Co., 1907
FAULKNER, R.O. *The Ancient Egyptian Pyramid Texts*. Aris & Phillips Ltd., Warminster, 1993 ed.
— *The Book of the Dead*. British Museum Press, London 1972
— 'The King and the Star-religion in the Pyramid Texts', in *JNES*, vol. XXV, 1966, pp. 153–61
FIX, WM. R. *Pyramid Odyssey*. Mercury Media Inc., Virginia 1978
FLINDERS PETRIE, W.M. 'The Building of a Pyramid', in *Ancient Egypt*, June 1930, part ii
— *The Pyramids And Temples of Gizeh*. Histories and Mysteries of Man Ltd. London, 1990 ed.
FRANKFORT, H. *Kingship and the Gods*. University of Chicago Press, 1978
— *Ancient Egyptian Religion*. Harper & Torch Books, 1961 ed.
GARDINER, A.H. 'The Secret Chambers of the Sanctuary of Thoth', in *JEA*, 11, 1925, pp. 2–5
GARDINER, A. *Egyptian Grammar*. Oxford University Press, 1957
GEESON, A.F. *Building Science: Structures*. The English University Press, 1967

GILBERT, A.G. (ed.) *The Cosmic Wisdom Beyond Astrology; towards a new gnosis of the stars*. Solos Press, 1991
GILLINGS, R.J. *Mathematics in the Time of the Pharaohs*. Dover Publications Inc., New York 1982
GOYON, G. *Le Secret des Batisseurs des Grandes Pyramides: Kheops*. Pygmalion, Paris, 1991 ed.
GREAVES, J. *Pyramidographia*. J. Bridley, London 1646.
GRIFFITHS, J.G. *The Origins of Osiris and his Cult*. E. J. Brill, Leiden 1980
GRINSELL, L.V. *Barrow, Pyramids and Tombs; Ancient Burial Customs in Ancient Egypt, the Mediterranean and the British Isles*. Thames & Hudson, London 1975
HASSAN, S. *Excavations At Giza*, vol. VI, part I. Government Press, Cairo 1946
HENBEST, N. *The Mysterious Universe*. Ebury Press, London 1981
HIRSHFELD, A. *Sky Catalogue* 2000.0, vol. 1. Cambridge University Press, 1982
HORNUNG, E. 'Die Kammern des Thot-Heiligtumes', in *ZAS*, band 100, 1973, pp. 33–6
HOWARD–VYSE, R.W. *Operations carried on in the Pyramids of Gizeh in 1837*, 3 vols. J. Fraser, London 1840–2
INGHAM, M.F. 'The Length of the Sothic Cycle', in *JEA*, 55, 1969, p. 36
ISLER, M. 'An Ancient Method of Finding and Extending Direction', in *JARCE*, XXVI, 1989, pp. 191–206
— 'The Gnomon in Ancient Egypt', in *JARCE* XXVIII, 1991, pp. 155–85
JONES, M. 'The Temple of Apis in Memphis', in *JEA*, 76, 1990, pp. 141–7
KANE, J.A. *The Ancient Building Science*. Edwards Bros. Inc., 1940
KERISEL, J. *La Pyramide à travers les Ages*. Presses Ponts et Chaussées, Paris 1991
KINGSLAND, W. *The Gnosis or Ancient Wisdom in the Christian Scriptures*. Solos Press, 1993
KRUPP, E.C. *In Search of Ancient Astronomies*. Chatto & Windus, London, 1981
LAUER, J.P. *Observations sur les Pyramides*. Imprimerie de l'Institut Français D'Archéologie Orientale, Cairo 1960
LECLANT, J. *Le Temps des Pyramides*. Editions Gallimard, Paris 1978
LEGON, J.A.R. 'A Ground Plan at Giza', in *DE*, 10, 1988, pp. 33–9
— 'The Giza Ground Plan and Sphinx', in *DE*, 14, 1989, pp. 53–61
LEHNER, M. *The Egyptian Heritage*, *ARE* Press, Virginia Beach 1974
— 'Some Observations on the Layout of the Khufu and Khafra Pyramids', in *JARCE*, XX, 1983, pp. 7–21
— 'The Development of the Giza Necropolis: the Khufu Project', in *MDAIK*, 41, 1985, pp. 109–43
— 'The Giza Mapping Project: Season 1984–1985', in *JARCE*, Newsletter 131, 1985, pp. 23–45

— 'The Giza Mapping Project: Season 1986', in *JARCE*, Newsletter 135, 1986, pp. 29–48
LETHABY, W. *Architecture, Mysticism and Myth*, (Foreword by A. G. Gilbert). Solos Press, 1993 ed.
LICHTHEIM, M. *Ancient Egyptian Literature, Vol. 1, The Old and Middle Kingdom*'. University of California Press, Los Angeles, 1975 ed.
LOCKYER, J.N. *The Dawn of Astronomy*. Macmillan, London 1894
LUCAS, A. *Ancient Egyptian Materials and Industries*. Histories and Mysteries of Man Ltd. London, 1989 ed.
— 'Were the pyramids painted?' in *Antiquity*, vol. xii, 1938
MALEK, J. *In the Shadow of the Pyramids*. Orbis, London 1986
MANCHIP WHITE, J.E. *Ancient Egypt, its Culture and History*. George Allen & Unwin Ltd. London, 1970 ed.
MASPERO, G. 'La Pyramide du Roi Teti', in *Rec.Trav.*, vol. v, Fasc. I–II, p. 1
— 'La Pyramide du Roi Pepi I', in *Rec.Trav.*, vol. v, p. 157
— 'Sur le Pyramidion D'Amenemhait III à Dashour', in *Ann.Serv.Tom.*, iii, pp. 206–8
MAYASSIS, S. *Mystères et Initiations de l'Egypte Ancienne*. B.A.O.A., Athens 1956
MCNAIR, W.A. *Starland of the South*. Angus & Robertson, Sydney 1950
MEAD, G.R.S. *The Doctrine of the Subtle Body in the Western Tradition* (Foreword by A. G. Gilbert). Solos Press, 1993 ed.
MENDELSSOHN, K. *The Riddle of the Pyramids*. Thames & Hudson, London 1974
MERCER, S.A.B. *The Pyramid Texts in Translation and Commentaries*. Vols. 1, 2, 3 and 4. New York 1952
— *The Religion Of Ancient Egypt*. London, 1946
MILLS, H.R. *Positional Astronomy and Astro-Navigation made Easy*. Stanley Thornes Ltd., 1978
MONTET, P. *Géographie de l'Egypte Ancienne*. Imprimerie Nationale, Paris 1957
— *Isis: ou à la Recherche de l'Egypte Ensevelie*. Hachette edition, Paris 1956
MOORE, P. *Guide to the Stars*. Lutterworth Press, London 1974
— *The Story of Astronomy*. Macdonald and Jane's, London, 1974 ed.
NEUGEBAUER, O. *A History Of Ancient Mathematical Astronomy*. Springer-Verlag, 1975
NEUGEBAUER, O. & PARKER, R. *Egyptian Astronomical Texts*. Vols. 1, 2 and 3. Brown University Press, Lund Humphries, London 1964
NOAKES, A. *Cleopatra's Needles*. H. F. & G. Wetherby Ltd., London 1962
OOSTERHOUT, G.W. 'The Heliacal Rising of Sirius', in *DE*, 24, 1992, pp. 73–111
OUSPENSKY, P.D. *A New Model of the Universe*. Routledge & Kegan Paul, 3rd ed., London 1937

PARKINSON, R.B. *Voices of Ancient Egypt, an Anthology of Middle Kingdom Writings*. British Museum Press, London 1991
PARMAN, A. & EL-SAID, I. *Geometric Concepts in Islamic Art*. World Of Islam Festival Publishing Co. Ltd., 1976
PIANKOFF, A. *The Pyramid of Unas, texts translated with commentary*. Bollingen, Series XL 5, Princeton University Press, Princeton NJ 1968
PIAZZI SMYTH, C. *The Great Pyramid: its secrets and mysteries revealed*. Bell Publishing Co. New York, 1990 ed.
RADICE, B. (advisory ed.) Herodotus. *The Histories*. Penguin Books, London, 1972 ed.
— Plato. *Timaeus & Critias*. Penguin Books, London, 1977 ed.
REISNER, G.A. *Mycerinus, the Temple of the Third Pyramid at Giza*. Harvard University Press, Mass., 1931
ROCCATI, A. *La Littérature Historique sous l'Ancien Empire Egyptien*. Les Editions du Cerf, Paris 1982
ROUSSEAU, J. 'Analyse Dimensionelle de la Pyramide de Chéops', in *DE*, 22, 1992, pp. 29–52
RUKL, A. (S. Dunlop ed.) *The Hamlyn Encyclopedia of Stars & Planets*. Hamlyn Publishing Group, London 1988
RUNDLE CLARK, R.T. *Myth and Symbol in Ancient Egypt*. Thames & Hudson, London, 1978 ed.
— *The Legend of the Phoenix, Part 1*. University Of Birmingham Press, 1949
SAGAN, C. *Cosmos*. Book Club Associates, London, 1981 ed.
SANFORD, J. *Observing the Constellations*. Mitchell Beazley International Ltd., London 1989
SANTILLANA, G. & VON DESCHEND, H. *Hamlet's Mill*. Gambit International, Boston 1969
SCHEEL, B. *Egyptian Metalworking and Tools*. Shire Egyptology, Aylesbury 1989
SCHWALLER DE LUBICZ, R.A. *Sacred Science, the King of Pharaonic Theocracy*. Inner Tradition International, New York, 1982 ed.
SCOTT, W. *Hermetica* (Foreword by A. G. Gilbert). Solos Press, 1992 ed.
SELLERS, J.B. *The Death of Gods in Ancient Egypt*. Penguin Books, London 1992
SHORTER, A.W. *The Egyptian Gods*. Routledge & Kegan Paul, London, 1983 ed.
SMART, W.M. *Text-Book on Spherical Astronomy*. Cambridge University Press, 1931
SPENCE, L. *Myths and Legends, Egypt*. Bracken Books, London, 1985 ed.
STADELMANN, R. *Die Ägyptischen Pyramiden*. Wissenschaftliche Buchgesellschaft, Darmstadt 1985
STEMMAN, R. *Mysteries of the Universe*. Bloomsbury Books, London 1991

TEMPLE, R.K.G. *The Sirius Mystery*. Sidgwick & Jackson, London, 1981 ed.
THOMAS, E. 'Air-Channels in the Great Pyramid', in *JEA*, 39, 1953, p. 113
THOMPSON, S.E. 'The Origin of the Pyramid Texts found on Middle Kingdom Saqqara Coffins', in *JEA*, 76, 1990, pp. 17–25
TOMPKINS, P. *Secrets of the Great Pyramid*. Allen Lane, London, 1973 ed.
TRIMBLE, V. 'Astronomical Investigations concerning the so-called Air-shafts of Cheops's Pyramid', in *MIOAWB*, band 10, 1964, pp. 183–7
WAINWRIGHT, G.A. 'Iron in Egypt', in *JEA*, 18, 1931, pp. 3–15
— 'Orion and the Great Star' in *JEA*, 21; 1936 pp. 45–6
— 'Some Celestial Associations of Min', in *JEA*, 21, 1936, pp. 152–71
— 'A Pair of Constellations', in *Studies for F.Ll. Griffith*, 1932
WALLIS-BUDGE, E.A. *An Egyptian Hieroglyphic Dictionary*, Vols. 1 and 2. Dover Publications Inc., New York, 1978 ed.
— *Egyptian Language*. Dover Publications Inc., New York, 1983 ed.
— *Osiris and The Egyptian Resurrection*, vol 1. Dover Publications Inc., New York 1973 ed.
— *The Book of the Egyptian Dead*. Dover Publications Inc., New York, 1967 ed.
— *The Egyptian Heaven and Hell*. Martin Hopkinson & Co., London 1925
— *The Mummy*. Collier Books, New York, 1972 ed.
WEIGALL, A. *A History of the Pharaohs*, vol. 1. Thorton Butterworth Ltd., London 1925
WOOD, H. *The Southern Sky*. Angus & Robertson, Sydney 1967
WOOD JARVIS, H. *Pharaoh to Farouk*. John Murray, London, 1955 ed.
WORTHAM, J.A. *British Egyptology 1549–1906*. David & Charles, Newton Abbot 1971
ZÄBA, Z. *L'Orientation Astronomique dans l'Ancienne Egypte et la Précession de l'Axe du Monde*. Prague 1953

Index

Page numbers in italics indicate illustrations and diagrams in the text